Ecology Works
John Newling

Beam Editions

Contents

Storm 2 – We need to be needed but forgot where we were, 2015
Lavender, ink, acrylic, fan, gilding

Foreword

Jonathan Watkins,
Director, Ikon, Birmingham

John Newling's new work exemplifies a strong environmental proposition – at a point in history in which humans are responsible for an unprecedented ecological threat to our environment, how should we now view our relationship to Nature? Involving a wide range of media, sometimes performative, often process-based and sculptural, it also pushes at the limits of what defines art – the gesture of giving gifts to passers-by, the act of nurturing and meticulously observing the growing of plants, the physical construction of soil from the printed word. And there is a bookishness of the best kind, manifested in publications that are integral to his practice overall. Preceding this one was *Dear Nature* (2018), a collection of 81 letters that he wrote *to* planet Earth, in a spirit, as writer Richard Davey states, '[in part] of truth and reconciliation, part advocacy of an urgent need, part thoughts for future social ecologies'. It is a remarkable project, as disingenuous as it is scientifically informed, as poetic as it is uncompromising in its conviction.

In one letter Newling quotes from the first chapter of Genesis, after Adam and Eve are introduced: 'God blessed them and said to them, "Be fruitful and increase in number; fill the earth and subdue it. Rule over the fish in the sea and the birds in the sky and over every living creature that moves on the ground."' And then he reflects on the implications:

> This is my guilt and worry. We have filled the earth. We have subdued it. We do rule over it.
>
> What is the cost to ourselves of subduing all that we seem to encounter?
>
> Yours is a vast history, ours is a blink in that history. Perhaps that's where the hope lies in our geological youth; we are young and still learning.
>
> I am sorry.
>
> Yours,
> John

The letters are matched by a succession of photographic images of a growing flax seedling – incidentally, one of the hungriest of plants – cultivated by the artist in his studio. Thus, we witness a miracle of life while reading about our downward spiral into the 'strangest nihilism without purpose and feeling'.

Including such a philosophical twist is characteristic of Newling, an artist who belongs to a generation that grew up with Conceptual Art,

Land Art and Arte Povera in the 1960s, and emphasises the importance of concept and context to his work. In 1985 he spent time in America as the first recipient of a Fulbright fellowship in visual arts. During this time he produced works of art in hotels, swimming pools, burnt out cars, at auction house sales of memorabilia, and on the streets of Los Angeles, San Francisco, Washington DC and New York. This varied experience marked the beginning of an interest in the meaning of 'place', something that informs his work now more strongly than ever, his place being a domestic back garden as much as it is the whole world.

In 2016 Newling developed a new alphabet based on leaves he found at Nymans, a National Trust garden in West Sussex. Once the leaves had been pressed and photographed, Newling worked with designers Jonathan Casciani and Oliver Wood of Beam Editions to create a typeface that matched these plant forms to letters in the modern English alphabet – thus, 'Nymans Language' was born, an officially registered, downloadable font and the subject of another book published by Beam.

When displayed in exhibitions, *Nymans Language* occurs in a grid of marble slabs on the floor, reminiscent at once of teaching aids for children and a Carl Andre sculpture, alongside a printed translation of the Genesis verses.

There are a number of new works that have involved the artist making soil, embodying text in various ways. *Waste Land Soil (notes)* (2018), for example, involved the shredding of several hundred A4 printouts of T.S. Eliot's poem 'The Waste Land', combined with coffee grounds and other kitchen scraps to create a soil capable of sustaining vegetative growth – the opposite of waste. Then, in a reverse process, Newling used the soil to make paper for his *Eliot's Note Books*: 'In this manner each page of each notebook contains the words of 'The Waste Land'. The notebooks evolve as a line of small sculptures, implying an imagined line of editing. There are 434 paper sheets in the work reflecting the 434 lines in Eliot's poem. The final books of the work are entirely constructed in a soil that, if placed in the earth, would grow and sustain life.' Newling's choice of 'The Waste Land' is very particular, as he explains:

> To make a project in response to Eliot's great poem is a privilege and a responsibility… It seems, at one and the same time, to fold many of our histories into the everyday; moments observed, and histories remembered. The poem constantly seems out of reach and mysterious, whilst giving just enough to keep our references to our own lives. It is a work that seems to possess a geology that layers the temporal, corporal and spiritual into its soil. A soil that speaks of a great poet/anthropologist of all times and none.

Newling has also been making *Soil Books* using soil from his back garden to convey the idea of the Anthropocene, our current geological age in which the human impact upon the Earth is irrefutable. He refers to us 'signing our names in the soil' (a phrase which appears often in *Dear Nature*) and is determined to grasp 'what is it to know that we have affected the effects of nature/our environment?' Continuing, '[Once] we were scared of nature. We were quite fragile as a species… you can trace our evolution to a point where we subdued nature, but to our own cost, because we will make ourselves extinct. To have a geological era where you see human-induced carbon in soils is quite extraordinary'.

The *Soil Books* are sculptures derived from Newling's everyday experience of walking from his house into his back garden. He picks up leaves along the way: 'It's like a ritual, so that every leaf in those books – the language of the books – is from my garden.' The result is book-like stacks of panels of processed soil, with the leaves pressed and gilded on the surfaces, which are then stained with watercolour – the colours chosen by the artist to reflect his feelings about the weather and atmosphere generally: 'Misty pinks at beginning of the year, through to dark greys/blacks at the end.' For this reason the order of the books is crucial, indicating seasonal change, and once again conveying the artist's preoccupation with physical transformation: 'I'm interested in things going from one state to the next state to the next state … never being quite sure what the end state is.'

When Newling paints onto soil, or makes marks on leaves, he is referring to human intervention in the natural world. Every work reproduced in this book is at once an expression of wonder at the everyday miracles that occur in the natural world, whilst also chiming with the kind of no-nonsense realism that James Lovelock espouses. Like Lovelock, Newling subscribes to the Gaia principle, understanding that as we sign ourselves into the soil, we are erasing ourselves: 'The earth does not care for us. If we leave the planet, the whole system of the earth will carry on very happily without us. It's so obvious.' His artistic practice is at once ritualistic – embracing ideas of the resonances of things past and passing – and completely in tune with a modernist aspiration to a truth to materials. As he says, 'It wouldn't work otherwise'.

Jonathan Watkins

Preface

Jonathan Casciani,
Director, Beam Editions

Ecology Works is a chronological account of John Newling's artworks, exhibitions, installations and artist's books made since the seminal project *Chatham Vines* in 2004. The book is not an exhaustive record of every work produced in this period, but is a record of landmark projects that have made a major contribution to the direction of the artist's work. Some of these are small, intimate projects that the artist has realised in his home studio and garden and others are major exhibitions and commissions.

The book opens with a series of drawings made during Newling's formative early years as an artist, *9 Twists of Nature* (1978–1979) (p 10). While this is an anomaly within the timeline, it is a key series of works that demonstrates the artist's early dialogues with 'nature'. Only one of the drawings has been previously published and we felt that *Ecology Works* was the correct context to show more of this work.

The book's opening essay *From My Garden* was written in 2004, a seminal work for the artist that predates much writing on art and the Anthropocene epoch. The essay was first published as a free newspaper that Newling distributed for free on the streets of Lincoln, UK, as part of the exhibition *The Noah Laboratory* (p 36). The essay is followed by *Art in co-relationship with nature* by Anne Douglas and Mark Hope who have been in close dialogue with the artist since 2013 and since Newling's residency at The Barn in Banchory, Aberdeenshire in 2016.

The third essay is written by author Richard Davey who has written extensively on the artist and his work since the early 2000s. His book *Spinning – Nature, Culture and the Spiritual in the Work of John Newling* was the first monograph on the artist that covers many works throughout Newling's career prior to his pivotal 2013 exhibition *Ecologies of Value* at Nottingham Contemporary (p 62). This book precedes *Ecology Works* and was an essential point of reference for determining the editorial frame of the book.

Curator and Director of Ikon Gallery, Jonathan Watkins, has kindly contributed a foreword as the publication of the book coincides with a major exhibition at Ikon, in the artist's birthplace of Birmingham. Collaborating with Ikon has enabled us to include much of the work from this extraordinary exhibition.

Alongside many of the artworks and projects in this book are descriptive extended captions, as well as extracts from the artist's own writings. Newling frequently writes short pieces of text when making new artworks to clarify his own thoughts. Sometimes these texts have appeared on the artist's website and in his seasonal newsletters and sometimes these have evolved into full essays and even books. Often they are given to curators to help form the interpretation for exhibitions. Where we thought necessary, we've written new extended captions in close consultation with the artist to create a more nuanced interpretation of the artworks to be read alongside and in contrast to the artist's own writing. In a sense, these extended captions are a collaborative text and exchange of ideas between Beam Editions, John Newling and his many collaborators since 2004.

As a close collaborator of Newling's since 2005, and having worked on many of the artist's books, I've overseen this process with my partner Oliver Wood to ensure these captions are insightful and accurate, while leaving enough room for the reader to create their own interpretations and be inspired by the work of this extraordinary artist.

Jonathan Casciani

Research Drawings, 9 Twists of Nature
1978–1979

9 Twists of Nature is one of the earliest examples of Newling's dialogue with Nature. Produced during his time at Chelsea School of Art, this series of 16 drawings imagine the construction, DNA and patterns to be found in trees. This is the earliest example of the artist exploring human relationships with nature and the connections between ecology and art.

While the drawings are entirely poetic, their analytical quality is a reference to scientific processes. Newling, throughout his 45-year career has continually explored the tensions between philosophy, science and morality. The artist has continually advocated for a society built on new values, recognising nature as central tenet.

The drawings are graphite pencil on textured art paper set on a hand-drawn grid with a stamp in the corner that reads 'Research drawings, 9 Twists of Nature, John Newling 1978–1979'.

Research Drawings, 9 Twists of Nature,
1978–1979
Pencil and ink on paper

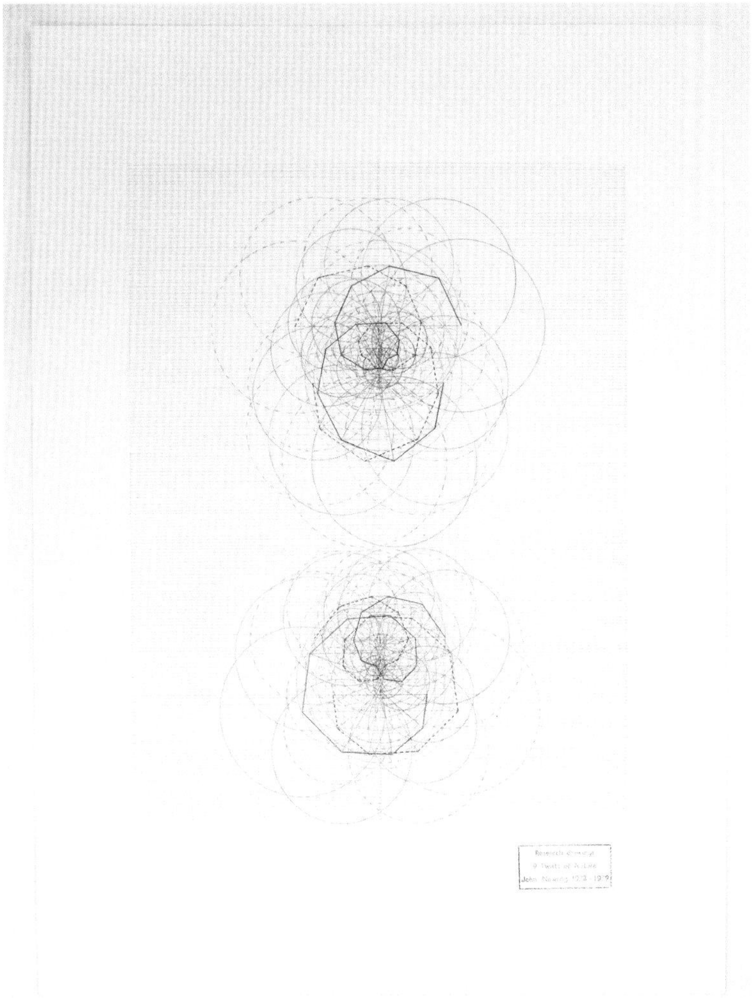

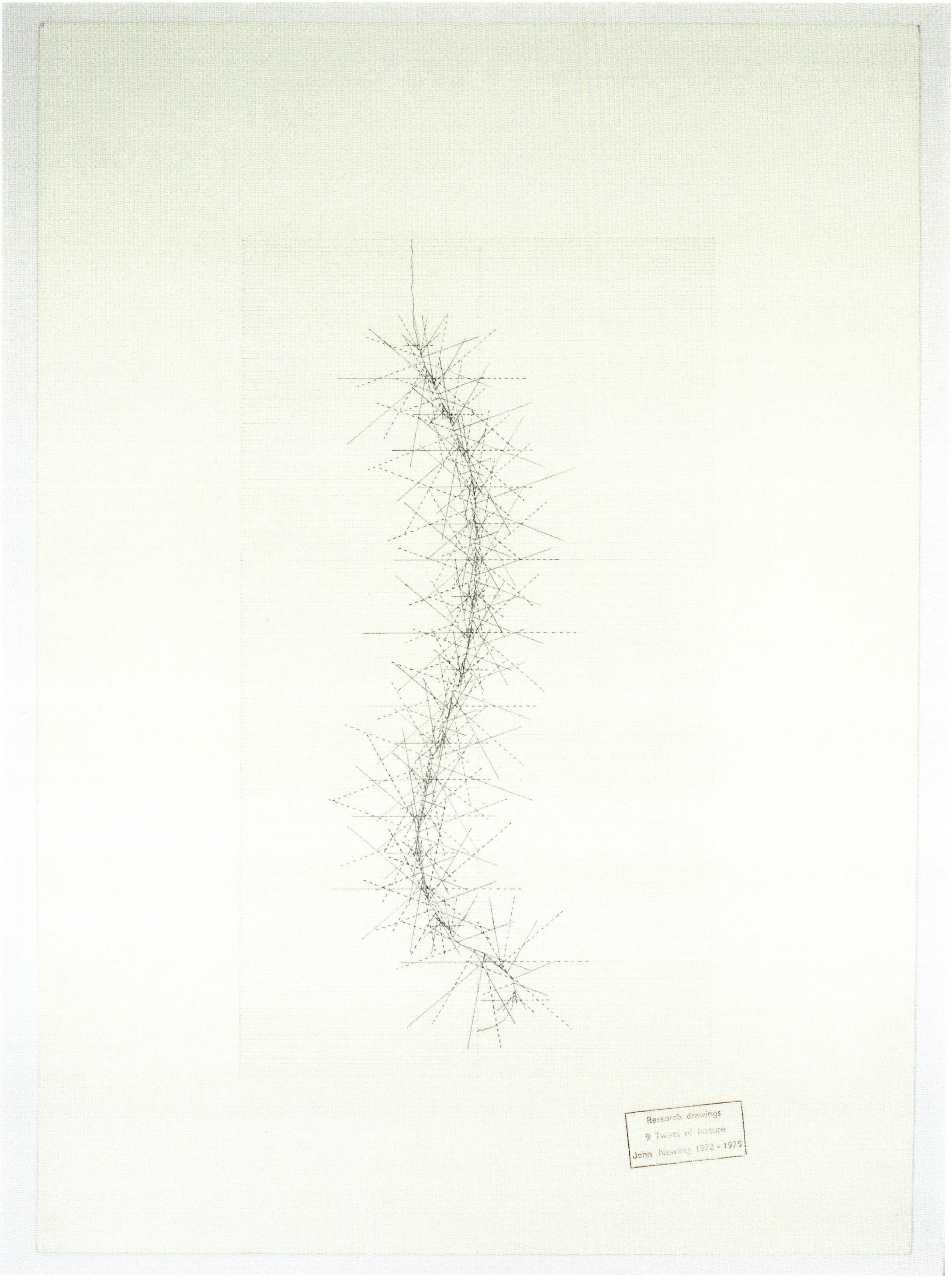
Research drawings
9 Twists of Nature
John Newling 1978 - 1979

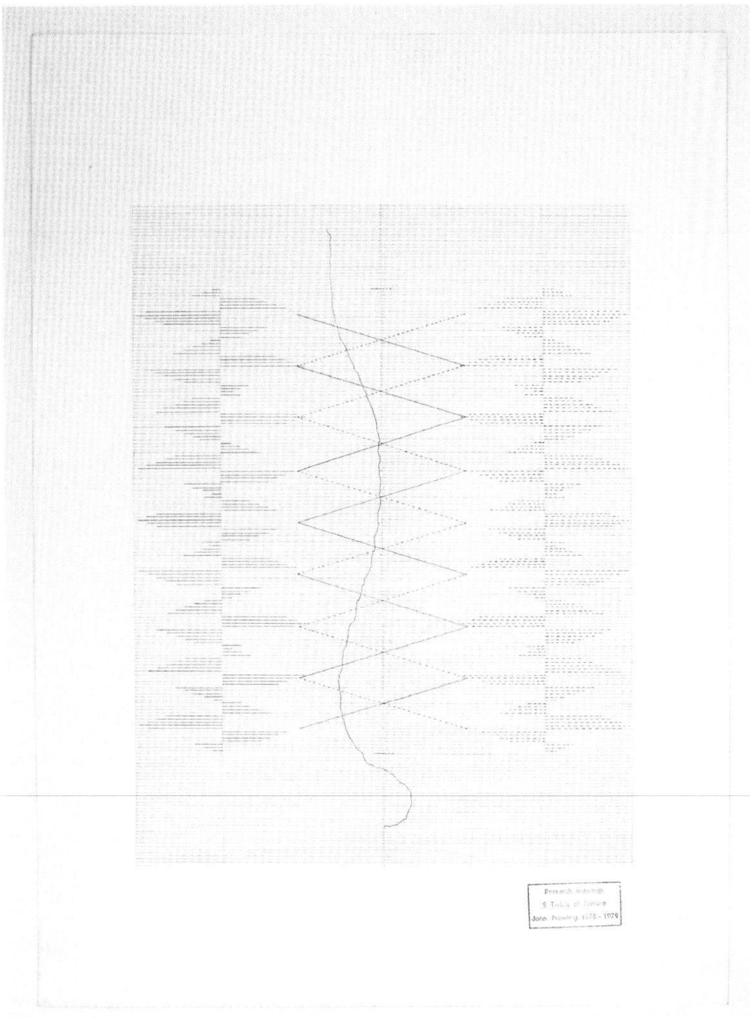

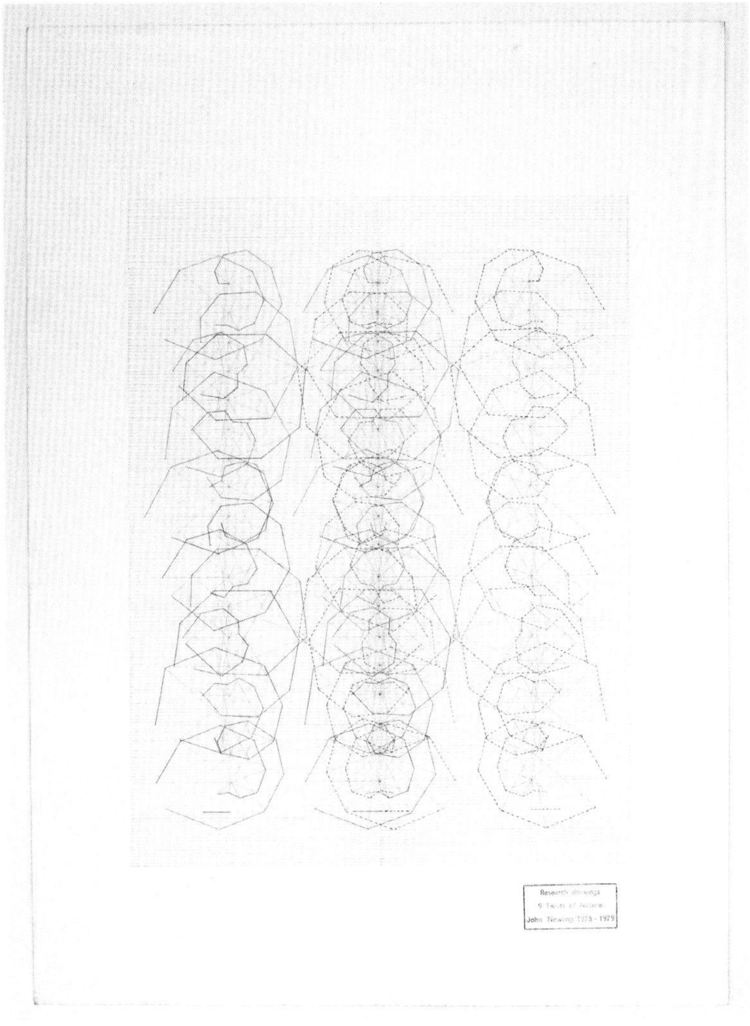

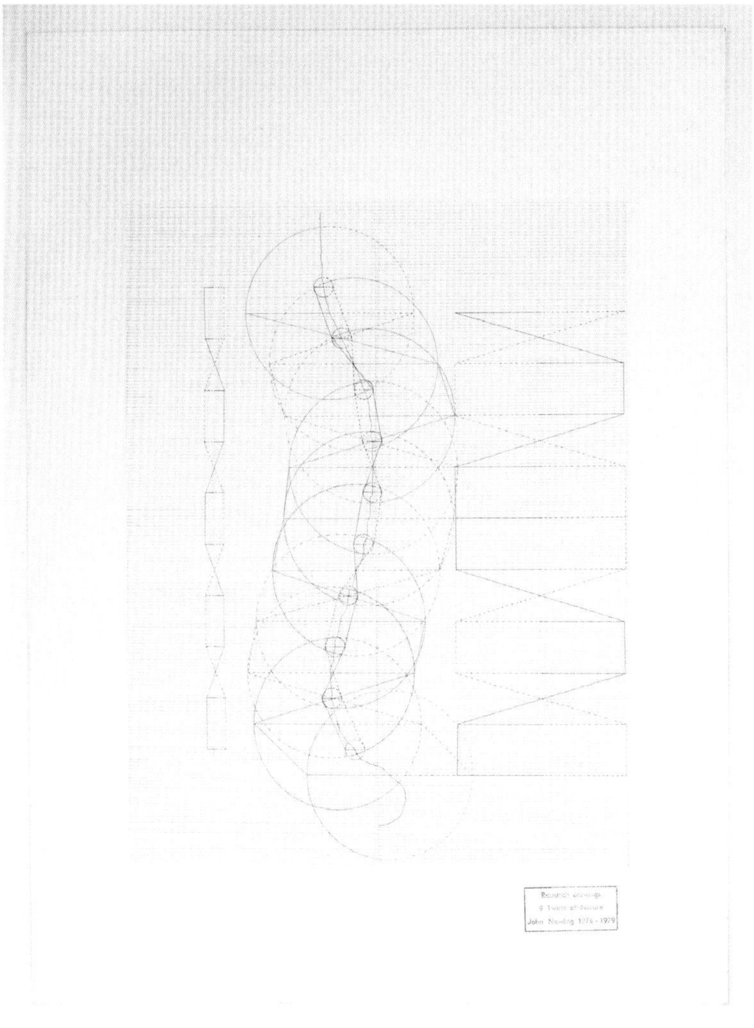

Research drawings
9 Twists of Nature
John Newling 1978 - 1979

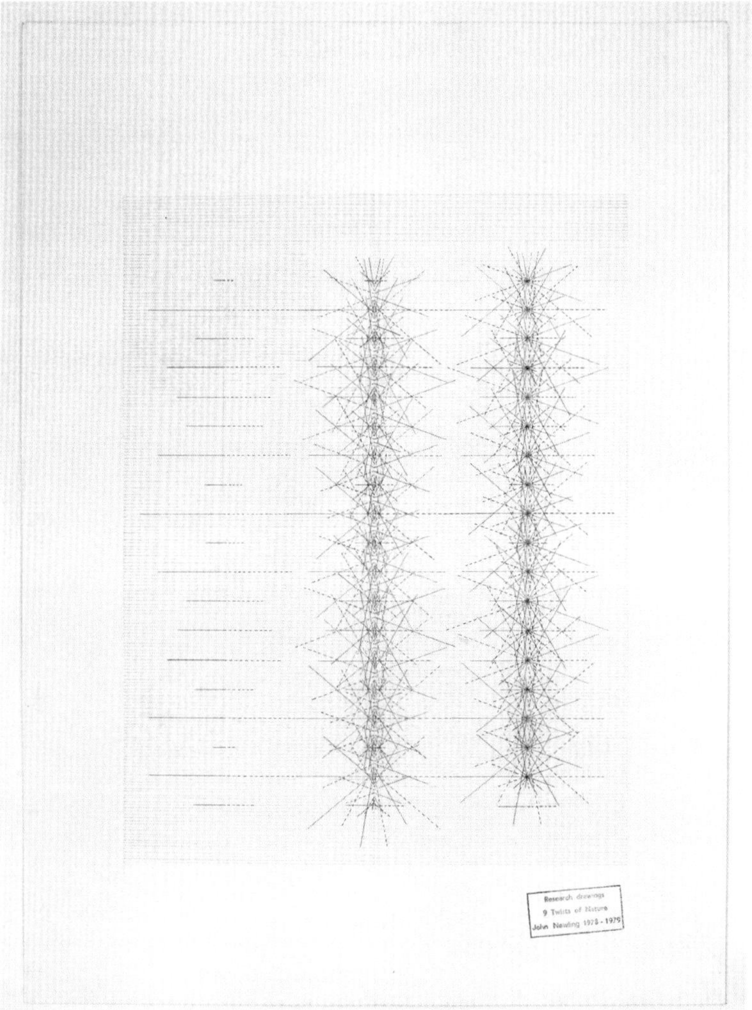
Research drawings
9 Twists of Nature
John Newling 1978 - 1979

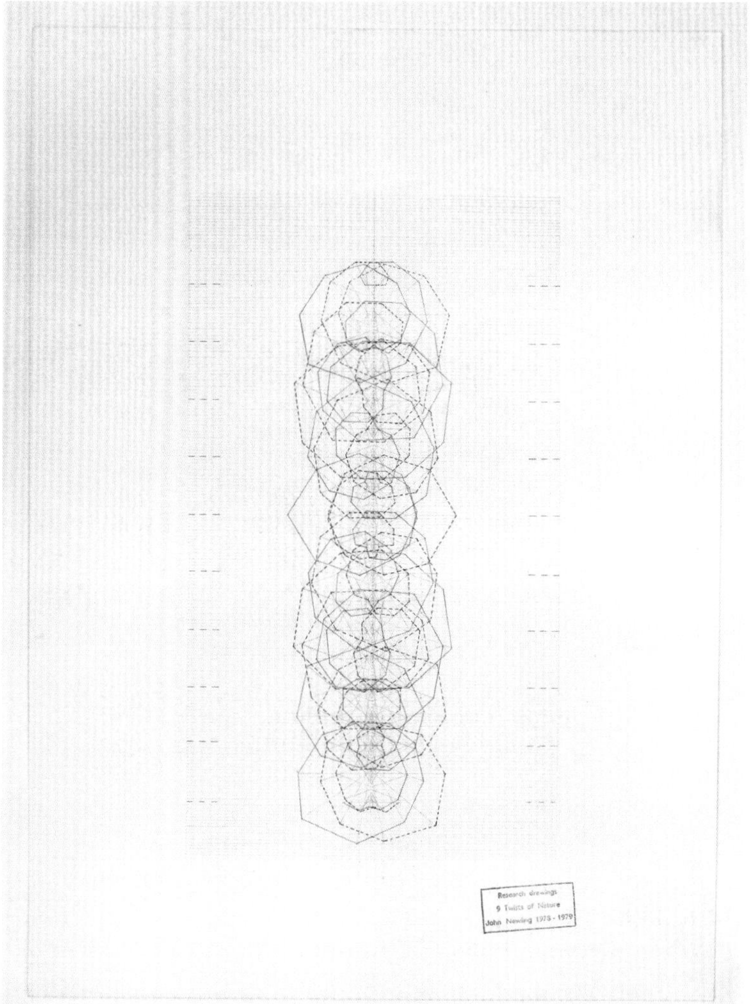
Research drawings
9 Twists of Nature
John Newling 1978 - 1979

From My Garden – Being Human in the Anthropocene Era

John Newling

In early spring 2008 I began to make notes on the ways we connect with our natural environment. In late summer I gathered these notes and started to draw together some of the reflections. Whilst I carried each thought around in me, my garden became the place for my reflections. A garden is a very particular type of clearing.

In our earliest days, engulfed in the murky dank of the forest, a clearing would have been a place where arboreal dimness would have been cut through with light. It would have been a place where the sky seemed nearer and the earth less cluttered. It is in forest clearings that our ancestors may have begun to consider their relationship to the natural world that surrounded them.

The philosopher Heidegger described the notion of the clearing as a metaphysical space where entities other than us can emerge. Entities that are made visible by being brought into the light is a poetic idea. It is also useful in helping us to consider a human form of being that Heidegger refers to as Dasein or 'existence'. The clearing is the place where entities that are independent of human purpose and human existence, seem to reveal themselves and fuse with human nature.

The clearing is any space that encourages the speculation and questioning that can connect us to the other entities that define our human-ness. Nowadays our clearings are not limited to the forest. The contemporary clearing is, in its function, the same as our ancestors but it differs in the multiplicity of places that these spaces can occur.

Changing the interior decoration of a house establishes a sense of personal ownership. It is a kind of removal of history. The garden, however, seems to evolve more slowly through its histories than the interior of the house. The transformations of the garden lend themselves to a deeper set of reflections and thoughts.

Looking back over the development of my thoughts the idea that 'Nature is independent of human purpose and existence' reveals an apparent chasm between views of the cosmos. On one hand is the idea of 'immanence'. This is the view that the cosmos has a mind within the world. An alternate view is to see the cosmos as a kind of machine, made up of working parts that are separate from our subjective selves. Much of this essay weaves its way in and out of this apparent divide, and questions whether it is wise to perpetuate the separation. Other thoughts explore knowledge, forged from complexity, which reveal an intricate network of ideas that illustrate our need to understand our connections to Nature. From religious practice to social Darwinism, this essay attempts to outline and review the inherent dangers in some of our methods of connecting to Nature.

At this moment in the early 21st century we are conscious that our human nature has affected Nature in a manner few predicted. This relationship has been based for millennia on the idea that Nature is something 'other' than ourselves. We have been shifting away from the 'other', that necessary and assuring machine, that enables our survival and yet generated a myriad of spiritual rituals in our need to communicate. We are now moving towards a consciousness of our effects on Nature.

The effect of this affecting has been to make us feel like a bad mechanic; helpless, guilty and confused. We can see that our human nature is capable of affecting a Nature that is an evolutionary determinant. We can also see that Nature does not need our causality. We are tiny within the universe. The ego of human nature is at once gratified and deflated. We have affected Nature, a control of sorts, but in doing so we have threatened ourselves. We are in a state of confusion, a kind of loneliness, in our relationship with the world.

I have become aware of the concept of a new geological era. The 'Anthropocene Era' was first used as a description in 2000 by Paul Crutzen, a Nobel Prize winning chemist. The term is used to describe the most recent geological era in the Earth's history where the soil itself gives evidence of the human impact on the Earth's climate and ecosystems. The Anthropocene era can be seen as emerging through the evolution and processes involved in settling. It is in settling that we begin to understand that we have affected the effects of Nature. I see the clearing as a place to better understand the relationship between humanity and the cosmos in the light of this new geology. In this essay I have considered the relationship through three distinct causal relationships. I have reflected on our relationship to Nature, to borders and to other species. I also attempt to review our contemporary soil from metaphysical, natural and human science perspectives.

This text emerges from contexts that help me to develop thoughts, questions and connections. As such, it is more aligned to human than natural science. There is a difference in the methodologies adopted by natural and human sciences. The natural sciences, like physics and chemistry, look towards a unifying theory of materiality using empirical methods. It is a theory that demands de-contextualization and subsequent abstraction. This is to say that a theory must hold true for all places and times to become a law. In the natural sciences, this has evolved de-contextualised entities. These entities, or thoughts, reveal themselves as independent of human purpose and human existence. They exist outside the necessity to correspond them to any given context and, as such, can be developed as law.

Human sciences, like psychology and sociology, cannot de-contextualize their observations, because they are largely concerned with us. In many ways the material of human science is human context. To advocate a theory of human nature that holds true for all places and all times seems unlikely given the peculiarities of human transactions and behaviours. The statement 'Nature is independent of human purpose and existence' must be subject to particular human contexts for the human scientist. Human views of Nature as either machine or as immanent cannot hold true for all places and times because of differing human views.

This essay is a review, from my clearing, of the confusion we have in regard to causes, connections and relationships we have with the natural environment. I share in this confusion and will continue to share in it but the reflection has helped me to clarify some of the elements held within the confusion. As such, my confused self is a little clearer and less lonely in the cosmos.

Our Garden

A resurgent set of thoughts

Over the years I have found gardening becoming a pleasure. This may be simply because I'm getting older and work with living things to better understand mortality. Whatever the reason, it is a pleasure to see plants grow and feel that I have been helpful in this process. I am not an expert and tend to go with instincts rather than knowledge. I am aware that most plants probably thrive anyway, despite my attempts to affect their wellbeing. I also know that if a plant does not respond to my attentions and dies, then I feel real remorse. It may well be that this desire to feel part of this urban nature is linked to a human quest to control Nature. In moments of reflection, considering that even if all humanity died the garden would keep growing, I am touching on surprisingly resurgent thinking about human nature in relation to Nature. Whether it is in a garden, a forest or a park, clearings are important to us all.

Our garden is a typical urban back garden. It has a shed at the bottom right hand corner and an oblong area of grass and moss surrounded by borders and hedges. At the house end, the garden has raised areas of rockery. In the bottom left hand corner there is an old apple tree. The borders of the garden are framed by ivy and privet hedges intermingled with several mature trees and shrubs. In front of the shed is a compost barrel. The small front garden is concreted and has plants in pots and a jasmine that is being trained to climb up the wall of the house. There are also plants in the porch. Hanging from the tree nearest to the kitchen window are several bird feeders. We are aware that foxes and other animals regularly visit the back garden.

Since moving into the house some sixteen years ago, Ann and I have made changes to the garden. This has been a very gradual process of adding plants, borders and a rockery. Each time we have planted something new, or dug a small border, we have had pleasure in watching the plants grow. On many occasions our planting has not worked and the plants died. Efforts are made to help the plant recover, attending to soil composition and watering, but it doesn't always work. Sometimes these familiar gardening activities provide me with a space where I can begin to explore the hidden-ness of Nature.

Changing the interior decoration of the house removes the history of those that lived there before us and makes the house our own. The garden evolves more slowly. Whilst the general architecture has not been radically altered, we have included plants that have changed the appearance of the garden. Unlike the transformations inside our house, those in the garden do seem to generate a deeper set of thoughts. The three most common subjects of my reflections have been our relationship with Nature, with borders and with other species. These reflections find expression within the clearing as illustrations of reality and the nature of being. A kind of ontological soil. This is an attempt to look at that soil.

Propositions from the garden

In the garden, I am aware that Nature is independent of human purpose and existence. In some ways this is amplified by thinking that 'If all humanity died – the garden would keep growing'. The garden, however, would be different if my purposeful changes had not been brought in to effect. This relationship between affect and effect within the clearing is at the core of understanding ourselves and other entities. The common thought, that 'If all humanity died – the garden would keep growing' reveals the idea of entities that are independent of us. It is an acknowledgement of our essential smallness within the larger view of our Nature, with Nature being an entity that is independent of human existence and purpose.

These thoughts appear in Heidegger's clearing. The human being and the being of Nature are contiguous, they are neighbours. Yet they are also essentially linked because human beings are dependent on Nature. This is evident in 'If I leave the garden alone it will keep growing', and also in 'If all humanity dies the garden will keep growing'. This proposition is incorrigible, that is, it seems true and is made in that truth.

Nature may be independent of the purpose and existence of humankind, but humanity is absolutely dependent on Nature. The existence of Nature does so, whilst Nature can be seen as being independent of us, we are entirely dependent on Nature. In the garden or on the farm this is evident in bad weather or plagues of insects or diseases. Gardeners can resent of such impositions realising that Nature is independent of their purpose. The garden will survive even if we do not. It is very human to attempt to understand Nature as a neighbouring entity within which we exist without being exclusive to its purpose.

It should not be forgotten that the frustrated gardener can also become the homeless farmer or grieving human as Nature affects us. Humanity is driven to understand the independent entity, Nature, in order to gain knowledge of its mechanisms. If we understand how Nature works we may aspire to some control. It is clear that, as a species, we have sought to commune with and control Nature as our cosmological friend. We know that we are dependent on this 'other' for our survival. Human knowledge of our dependency on Nature has generated rites and beliefs that try to influence our environment.

A huge and rich anthropological soup of rites has been formed to help us feel as if we could influence Nature. Within these rites there are views of Nature as a machine, a set of causes and effects that are separate from us. But there are also views of Nature as an immanence held within all things including us.

In the garden, my pleasure in seeing a plant thrive is, in a small way, a feeling that I can control Nature. That is, I can affect its effects. This would lead to the proposition, 'Gardening is independent of Nature's purpose and existence'. Well in some ways this may be true. Gardening as a contextually specific activity may well not be part of Nature's purpose and existence. Gardening seems to be an opportunity for human beings to experience tiny acts of causality within a highly organised and aesthetic situation. It would be difficult to imagine that gardening was part of Nature's purpose; it is not a warranted proposition, not a true belief based on knowledge. We know that without Nature the garden does not exist but the reverse is not true. The lack of a garden here or there does not affect Nature. Whilst the purpose of Nature is not to make my garden look beautiful, without Nature there would be no garden. Causality, here, is one of the small interventions in the existence of Nature, that independent entity whose purpose we struggle to know. Gardening makes my activities contiguous to Nature and my garden contingent on Nature.

When I'm working in my garden I see the small acts of assistance I render through two particular frames of thought; the mechanical and the immanent. The mechanical view is that Nature is a mindless machine whose working parts need to be understood in order to assert control over it. Taking the mechanical view means that I can consider the death of my plant is like breaking part of the machine. This makes me the incompetent mechanic that needs to understand more about the mechanisms involved to make it work properly.

Another view is to see Nature as immanent. Immanence derives from the Latin *in manere* 'to remain within'. Immanence refers to philosophical and metaphysical theories of the Divine existing and acting within the mind, within the world. This is a view that advocates that Nature is not independent of human purpose and human existence; instead Nature is inextricably linked through the Divine acting within the mind of the world.

The idea of 'purpose' here is a key to understanding a cosmos that has meaning. This is transcendent Nature. Throughout history, humans have attempted to give Nature a human purpose through notions of immanence. Many religions share this idea, that the cosmos has a purpose for us beyond the separate unthinking entities of mechanical Nature.

Our view of our natural environment is based on Nature being either immanent or mechanical. This is to say that the patterns we have learnt in relation to weather and season and other naturally occurring events are mechanical; we predict they will happen according to our experience and knowledge. Alternatively these events are immanent, caused through an agency that acts within the mind of the world as a whole.

When Noah started again he must surely have thought of Nature as independent of human purpose and human existence, a particularly lonely thought. He too, may have concluded that humans along with all living things were both contiguous to Nature and contingent on Nature. He may have had a sense of Nature as immanent, a Nature that acts within the mind of the cosmos. Noah would have called this God. Whatever thoughts he may have had, it seems a good bet to claim he would have thought of affect and effect.

A plant dies in the garden

To think I have affected the effects in my garden is a threshold to the clearing; it is a contextual sharing of a situation that is dependent on Nature but occasionally shows human involvement. It is a feeling of sharing causation within Nature, but human nature is dependent and not causal. These small acts of gardening invite us to the threshold of the clearing; they introduce a context where human nature seems to affect Nature. Such acts of causation are the thresholds of the clearing.

We garden despite knowing that if all humanity died the garden would keep growing. We garden to see that the small acts of causation we have learnt from Nature have an effect. This knowledge makes us feel less lonely in the face of knowing that Nature is independent of us.

We become aware that Nature is independent of us when the act of gardening goes awry; those times when acts of Nature have effects that frustrate our own attempts at affecting Nature. It is at these moments that human nature enters the clearing to experience and learn more of the 'other'.

If my actions of feeding and improving the soil to encourage plant growth fail and the plant dies I experience a sort of anxiety. I have regret for the dead plant and become anxious as I realise that the garden is not really subject to my influences but is wholly subject to Nature. These moments, which I have always thought of as a 'wobbling' of our certainty, are the times when those other entities come into a human context. The death of a plant brings a remorse that forces human nature to recognise Nature.

Less catastrophic, perhaps, than the death of other things, the death of a plant touches on mortality in general. Death, the most extreme event in human nature, does open up a context for the fusing of human nature within Nature. These constantly occurring events create thresholds to enter the clearing. The thresholds nourish an understanding of the ontology of the entities that inhabit metaphysics.

Learning in the garden

The death of a plant makes me re-think my relationship to the Nature in the garden. I have to accept a new relationship that is based on a Nature that is independent of human purpose and existence. How, then, does this affect my gardening in the future?

This break down of understanding in one context to experience a de-contextualised truth is learning through experience. It is a phenomenological event in that the death of the plant appears to us in both experience and consciousness. It is also hermeneutic in that the experience seems to need interpretation to avoid misunderstanding. Interpretation repeats the learning. It emphasises that my garden is not entirely governed by me, that the plant died anyway despite my best efforts. After considering what I could have done better, the new truth becomes apparent. My plants and I are both subject to the Nature that is independent of human nature. My best efforts are reflected in some understanding of Nature, but I cannot cause Nature, I can only mimic and repeat its effects.

However, this learnt repetition is truly amazing. Over millennia we have used our senses to interpret our experiences. This hermeneutical, phenomenological approach to learning has evolved our understanding and relationship to Nature.

Borders

Borders in the garden and elsewhere

I bought a border spade the other day. The adjective 'border' made me think about the function and consequences of the spade within the Anthropocene era.

The natural environment forms natural borders like mountains, rivers and coasts. We are also familiar with the human borders that cut across the landscape, borders that are political boundaries for local or national governments. The configuration of these borders change as war and politics change the geo-political map of the world. There is a cartographic skin of humankind lying on the surface of the natural landscape.

On either side of a constructed border the soil may be identical in composition, but what is grown in that soil is known by its national origin. If, for example, a vineyard were grown across a national border the wine produced would share provenance or origin through the soil. However, when sold, the wine would be labelled with its origin described by nation state. The border identified on the label influences our opinion on the wine. It delineates a human context rather than the natural environment because it frames the boundaries of a nation not a landscape. This is to see human constructions overlying natural law; one is to do with context the other is de-contextualised, common to all places and times.

When a new border is cut and dug in the garden it alters the architecture of the garden landscape and separates the border from other garden places and plants. Creating a border in a garden is an action of separating and segregating plants. The border is often given a particular function within the garden. The border then somehow moves away from de-contextualised natural laws towards a contextual view that assigns value through the specific subject of the border. The construction of a border in the garden begins to move us away from the proposition that 'The purpose of Nature is not to make my garden good yet without Nature there would be no garden'. It questions the proposition as a true belief based on knowledge.

Gardening is a pleasure, enjoyed by many, but it is probably more concerned with immanence than gaining knowledge from Nature as machine. Perhaps it is more spiritual than an essential means of production. An essential production that is intimately dependent on the natural environment is farming. The advent of farming may be the answer to the question: when did we stop belonging to a place and have the place belong to us? Farming creates borders, physical, social and psychological.

Analysis of Anthropocene geology suggests that farming has affected the environment. It seems inevitable that eventually we would affect our natural environment as human history shows our endeavours to do just that. Farming shifts our relationship as increasing knowledge allows us to control Nature. This shift is intensified by dividing general human understanding into specialist divisions of human subjects.

Borders or boundaries of knowledge can be compared to borders or boundaries of a field. Both are governed by our need to divide a subject so we can better control and understand the specifics. Subject borders have fostered the amazing growth of knowledge about the cosmos and much within it. They have also acted as the barrier that has separated and excluded many people from that knowledge and may now even threaten our environment.

The choices are stark. We can try to control and affect the natural environment with the knowledge and technology we have. We can also try to re-evaluate the way we view it in order to stabilise the relationship we have with Nature. This means removing not building barriers. At the very least borders should be crossed, subject disciplines less tightly bounded and knowledge shared across our species.

Our connections to the natural environment have been lessened by segregating land for food production. The irony of this action is self evident. Farming is an essential activity, it feeds us. But part of our present dilemma is that the very construction of land borders to intensively grow food excludes us from understanding the contiguous position we have with our natural environment.

Our relationship to the natural environment is contiguous whilst our existence is contingent on it. This is not to denigrate farming, an attitude popular with urban peoples for a couple of centuries. It is an attempt to connect the evolution of borders with our evolution of

knowledge acquisition. Both have functions that have contributed to our current situation.

The establishing of borders, both physical and psychological, may have shifted our view of the cosmos as a mind towards the idea of the cosmos as a machine. Both views seek correspondence with the effects of the cosmos. One view sees human purpose as being in a state of trust with a cosmos that must have a mind. The other sees human purpose being like a mechanics where the cosmos is a machine that can be dismantled to understand it. The truth seems to be somewhere between these views; both seem to be borders. Maybe it is possible to view the cosmos as both a machine and as a mind.

A transgression across the dominant borders would allow us to understand our purpose better within a cosmos of such wonder and complexity. It may also change our relationship to the cosmos if we thought of ourselves as both a part of the machine and the mind. Human purpose could then be seen as contiguous with the cosmos, part of its mechanics and its mind. Human science could become part of natural science as the human species is viewed as a component of the cosmos. As such, science could be seen as an element of both machine and mind; an element that can be de-contextualised in order to be true to all places and times.

In this way it becomes possible to advocate human laws within natural laws, seeing us as part of both cosmic machine and mind. However, it would still be true to say that if, as a species, we die, the garden would keep growing. This is an issue of existence rather than purpose.

In terms of our purpose, we seem predisposed to affect all the effects that we experience. By seeing ourselves as being part of the machine and also within the mind of the cosmos, we can understand that we can affect our environment. This creates a new kind of responsibility where we are not separated from our environment and we can begin to affect our natural environment in a way that benefits us all.

The Anthropocene era shows the changes in our relationship to the natural environment, but it may yet see us making changes through understanding that our relationship to the cosmos is that of being part of the machine and part of the mind.

Perhaps we simply do not know or understand what kind of machine or mind we are trying to connect to. We are a minuscule aspect of the cosmos but by advocating that we are part of its machine and mind we become less lonely in it. We better sense our belonging.

Crossing borders: consciousness to conscience

Many of the thoughts expressed in this text are reflections on the effect of humans on Nature. Attempts to understand the way we affect Nature is a struggle to cross the borders that these reflections may posit. Perceiving and crossing borders are both subject to consciousness of them. Both involve being something and being conscious of something. One concerns subjective experience, the other concerns information that can validate the subjective experience.

Ned Block, a contemporary philospher, has divided consciousness into phenomenal consciousness (being something) and access consciousness (being conscious of something). In the garden I am something and I am also conscious of something. The information necessary to be conscious of something is evident to my senses and is aquired through knowledge and experience. One consciousness could be described as natural, biological law of human-ness. The other is a social consciousness where contextual experiences generate information and knowledge.

In the garden I am conscious of my purpose; to encourage plant growth. I am also conscious that any growth is subject to natural laws that are independent of my purpose and existence. There are other species in the garden. They are also something and, possibly, conscious of something. My desire then is to understand and connect to the natural laws and also to speculate on the consciousness of the other species. I wish to cross the border from being simply contiguous to these other entities to feeling a greater sense of connectedness. One way of doing this is to acknowledge that consciousness is not exclusively ours. It may be a particular conceit to consider that, as a species, we are exclusive in our consciousness.

Across many cultures and times humans have endeavoured to connect with the natural environment. Through invention, imagination, curiosity and drugs humans have tried to expand their consciousness by sensing a greater consciousness in other entities. We have tried to cross the borders between them and us. The eagle, lion, storm, plant and other entities have all shown us attributes that we have desired, and at times we have sought to possess. By attempting to become something other than ourselves, we are attempting to connect to something deep within it. To become that life-form is to connect to that soul. This is an animistic view of other entities in our world. Animism, an ancient set of beliefs, sees everything in Nature as having a spirit or soul. Animism implies a consciousness in all living things; another example of us being lonely in the cosmos. This may be so, but the desire to connect is part of our spiritual welfare.

In the garden, my consciousness tries to cross the border between my being there to affect Nature and my being there to empathise with the effects of Nature. At the very least, the sense of being conscious of something other enhances a feeling of responsibility towards it. The nature of the responsibility is a spiritual welfare that is a progression from consciousness to conscience. An understanding of that informs our actions through our moral judgements. Attempts to cross borders between us and Nature can help to clarify our confusion as we experience being human in the Anthropocene era. This is a border crossing through 'phenomenal' consciousness to 'access' consciousness. It allows entrance to metaphysical, and actual, spaces where connections can be made between all living things and phenomena. A process that affects our conscience.

Both the physical and conceptual cartography of the garden has prompted me to consider borders in terms of species, peoples and nations. This has led to me thinking more generally about crossing borders. These are material borders; borders that divide the landscape made by human action. Informing some of these actions are borders of knowledge, more precisely, the ways in which we construct knowledge. To see the effect of these borders we need to consider how different disciplines may construct their own borders. These are the discipline borders used to organise, valuc and construct specific knowledge. Methods within and across borders.

All disciplines have subject borders. Having two or more disciplines applied to a single project is now common within many institutions both public and private. There are several types of inquiry that could be referred to as interdisciplinary. Multi, inter, trans and cross are all prefixes that describe the inclusion of another discipline.

In the garden, my thoughts are governed by crossing many subject borders. Thoughts are constructed using the knowledge of a discipline and the evidence of my senses. I deploy theories, methods and instincts to my thoughts to support my 'access consciousness' and organise my thinking. The subsequent knowledge that may evolve has many discplines within it.

Most important is the discipline of connection. It is a process of making subject borders more porous so connections between subjects can happen. This could be seen as a violation of other disciplines. I do not consider this to be so. My intention is simply to reflect on the way we experience events. In the garden I see that knowledge cannot have a single discipline as its origin. I see knowledge as a continual fusion of subjects informing each other like a constantly evolving, living construction; not unlike the plants that I have been observing. I believe these connections between and through subject knowledge are core to creativity.

The notion of knowledge being a living entity, constantly informing and re-informing our understanding of the world, can raise problems within institutions that audit the value of single subject disciplines. These institutions often assert that rigour is lost with intrusion from other disciplines.

The problem, of course, is not rigour. Rigour is desirable. The issue of rigour is linked with the problem of autonomy. Supporters of a single discipline will often argue that knowledge is a rational, informed and un-coerced entity that is put 'at risk' with the introduction of another discipline within its borders.

Subject disciplines can become unable or unwilling to disseminate knowledge to a wider public; us. The fear of losing self governance can lead to a lack of communication; yet loosening the borders of subject knowledge would create a wider democracy of knowledge.

We need to understand the implications and moral situations within the singularity of many subject disciplines. Much research is funded from the public purse and we are, therefore, part of it. If we cannot contextualise and connect that knowledge then we are being excluded from its significance. The borders of these complex subjects do need to attempt outward routes of dissemination so that we can all have an opportunity to connect to the current thinking within our world.

Part of our confusion is because we have little or no idea of what some disciplines research, why they do it and how it can be applied. We cannot reason their purpose or our involvement. The disciplines become institutionalised as they pursue rigour within tightly defined boundaries. This is a particular problem when trying to understand the relatively new sciences related to bio-technology. They are complex. In terms of public understanding what they do demands a high degree of specialist knowledge. However, understanding why the research is necessary may not be so difficult. The 'why' becomes the important ground in terms of dissemination and debate within a wider public. The 'why' is the important generator for our accommodation of knowledge. From my garden this particular 'why' has helped me better understand our relationship to the natural environment.

In the instance of trans-disciplinary work the notion is not to coerce or corrupt any single discipline but to attempt to create a situation where an investigation is informed through many disciplines. It has the structural distinction of removing or crossing discipline borders. Trans-disciplinary work establishes new insights and areas of research that are not necessarily quantified or qualified through a single discipline. Rigour and autonomy are not exclusive to the methods and structures of single discipline.

Knowledge in the clearing is Trans-disciplinary. The clearing generates complex structures that create and share knowledge in the understanding that knowledge does not have a single origin and is formed through a living process, where borders are transgressed and re-oriented.

Such an approach may loosen, rather than lessen, rigour in order to achieve new connections and approaches; an approach that crosses borders yet retains an autonomy can disseminate its purpose across many peoples.

Other Species

Animistic desires and the pet shop

Our garden is occupied by a variety of other species. Birds, foxes, squirrels, cats and bats are regular visitors. Ann and I get pleasure from observing them and have set up several bird feeders near our kitchen window so we can watch a variety of birds feeding. Foxes often wander across the garden and did have a den in the bushes near the house. Cats posture and hunt in the garden. Grey squirrels go giddy, jumping, climbing and digging in our garden. Bats do their dizzying dance at dusk as they feed. We live in a natural environment that has other species in it.

These visitors provide us with both amusement and wonder. Most importantly we see other living entities that feed our imaginations: what if I could fly like that? It leads us to speculate on their relationship to their natural environment. Inevitably we tend to anthropomorphise each animal, giving them characteristics of our own. A good example of this would be the bird boxes that I bought as gifts on Lincoln High Street. The boxes are shaped as a human type of house, pitched roof and all. Whilst this pays homage to the disneyfication of our relationship to other species it may also reveal a deeper need in us. We are lonely in our relationship to the natural world.

The human species has kept the natural world at a distance. Maybe this is because of our 'fight or flight' survival instincts or our 'learn and control' connections to Nature. The need for an 'other' in relation to this connection evolved the 'immanent' and 'transcendent' view of the cosmos and is important because it is part of our need to connect our existence and purpose to the natural world. Loneliness has played a part in all of this. When I see a fox or bird in the garden part of my pleasure is knowing that another species, with other ways of being and expression, exists along with me. My desire to relate to these other species, to further deepen my relationship to Nature, edges towards Animism.

Animism views everything in Nature as having a spirit or soul that is a part of the immanence of the cosmos. This would include waters, fires, rocks, plants, animals, humans and other elements of Nature. We are a tiny aspect of the cosmos, but by advocating that we are part of its machine and mind, we become less lonely in it. An element of such advocacy is our relationship to other species. To this end, the shamanistic practices of communicating with the spirit world have been developed for us to connect with and help us feel less lonely.

As I stand at our open kitchen door I wonder what that cat may be thinking or where these birds sleep. I ponder a whole heap of questions that attempt to connect me to them. In reconsidering the proposition 'If all humanity dies the garden will keep growing', as an essential truth, it is clearly a generative proposition that has encouraged us to try to communicate with all of Nature. Not least, the proposition encourages human attempts to speculate upon the idea of the spirit as never dying; the achievement of a kind of immortality as an element of both the mechanical and transcendent view of the cosmos. Parts of shamanistic practices are evidence of our hunger to see beyond our mortality, a border that has engaged us from our beginnings.

There is no doubt that pets are important in the society in which I live. A very considerable industry has grown around pets and our perception of their needs. Trips to a pet shop will reveal a thriving industry. Here the various necessities for pet comforts are laid out. Pet toys, beds and homes all pay testament to our desire for our pet to have a great life. The death of a pet is often met with undeniable grief. Pets have a dependency that lets us include other species in our desire not to be lonely in Nature and, in turn, with the cosmos. The pet shop becomes a metaphorical lonely hearts club.

These are domesticated animals whose existence is dependent on us. Unlike farm animals whose domestication has the function of feeding us, pet animals have a different function. They can make us feel part of the natural world. There is research that suggests pets are good for our wellbeing. In this post-industrial society they can foster a sense of trust in a world where our trust in so many things has been damaged. It is odd that we sometimes trust our pets more than our human relations. Maybe, as a species, we are unsure whether we are a pack or a lone animal.

Back in the garden the other species continue to occupy the space. Cats exhibit the familiar, individual, somewhat autonomous aloofness. Squirrels continue to dig, jump and scurry frenetically, most often with other squirrels. Birds continue to feed and fly. The bees seem to love the lavender bush and exhibit all the signs of intense and important work. Other species leave tracks and traces in the garden.

I recently met a man walking the street accompanied by the largest rabbit I have ever seen. The rabbit was on a lead and attracted much curiosity from passers by. It transpired the man bred these giant rabbits and was well known locally. As I watched the man and his rabbit talking with the groups of people that gathered round him, I could not help but speculate on our desire to communicate with other species as an element of trust in our relationship with Nature. In this instance, another species can be seen as both contiguous and contingent to human purpose and existence.

Constructing a Soil

The anthropogenic within the natural

Much of what I have written attempts to know the unknown. In some ways this is a metaphysical determinate acknowledging that there may be more to our understanding of our world than we know. Since our beginning, knowledge has been an essential tool in our survival. Without attempts to know things, including the knowing of possible unknowns, we would struggle to survive. The knowing of unknowns is a principle within the concept of the clearing. The 'known unknown' feeds our sense of detection. Such knowledge of the unknown is, essentially, a driving force for us. We would stop the necessary

challenges and quests to find out the unknowns and know them.

The architecture of the clearing is constructed with unknowns. We are lonely and lack a sense of our purpose beyond survival or existence. We have affected our natural environment and yet need to better understand our relationship with it. We have seen that this is a question of causality related to our existence and purpose. It should not be forgotten that we live in a world where the purpose of many is still simply survival. Our privilege surely gives us the purpose of helping to make a world where survival and purpose become less predicated on each other. The clearing is a space that helps us see our relation to the natural environment in terms of entities that seem to be independent of human existence and purpose.

The challenge of the Anthropocene era is to understand our existence so we can sense a purpose in our actions and thoughts. As an evolving geological era it would be good to think that the next hundred years would provide soil samples that indicate a reappraisal of our relationship to all things living; that the soil itself would tell the story of our better understanding of our relationship to Nature. The soil may begin to demonstrate that whilst Nature is independent of humankind we slowly began to understand that we are dependent within this independent entity. The ontology of such a metaphysical soil would be filtered through an understanding that the immanent and machine views of the cosmos were not binary and could coalesce as a combined understanding. A soil that inhabits souls and machines. A soil that could combine both the natural and human sciences and contain elements that are true to all places and times. Such a soil could, in the future of the metaphysical clearing, make us less lonely in the cosmos. A new purpose freed from existence.

Purpose in the soil

When a new plant is brought into the garden I am aware that my gardening purpose is to make a situation that is good for the plant. I am also aware that while I share with the plant the desire that it should thrive, the purpose of the plant may be different. The purpose of my gardening is socially engaged whilst the purpose of the plant is, as far as we know, entirely engaged with the natural world.

We know from Darwin's 'Origin of Species' that the struggle for survival is determined through natural selection. First published in 1859 this great work is an essential text to evolutionary biology. In the garden I separate my purpose to foster the plant from my existence, my need to survive. My purpose in gardening is not to ensure my survival. This is in contrast to the plant's purpose which is wholly predicated on survival. Given natural selection, one could argue that learning rudimentary aspects of plant welfare is a trait that my species should perpetuate. Clearly, it would be curious if our evolutionary selection was by a genetic predisposition to make good gardeners. Yet all our experiences make us selectable in order to perpetuate our species.

The theory of evolution describes survival through maintaining and developing essential traits. In the Anthropocene era, where we can see that we have affected our natural environment, it may be that our survival does become our purpose.

This would require a more focused view of our relationship to that environment. As a species we have exceptional gifts of detection and imagination that we can deploy on any given problem. My gardening may not have a distinct evolutionary purpose, except that it gives me pleasure. But it can be argued that, as an activity, it involves detection, imagination and evidence of the senses; qualities that strengthen our reasoning and emotional abilities

For our species, and maybe others, the relationship between purpose and existence is not entirely set. My purpose is not consciously linked to my existence or survival; as a gardener, my purpose is largely to encourage a plant to grow. My concern is with the plant's survival not mine. My connection to the plant can be seen as causal, in the knowledge that my best efforts may be thwarted by the real causality of Nature. Darwin, and subsequent evolutionary biologists, tell us that we are all connected. The causality of our natural environment affects the very characteristics that determine what is genetically valued in evolution. So, when we affect our natural environment we also affect the evolutionary characteristics of all living things. This connectedness is at the core of the Anthropocene era.

As climates change living entities need to adapt to that change. It may be an evolutionary predisposition to achieve an awareness of our influences on the environment and the consequent effects on evolutionary characteristics.

Parallel to this realisation is the rapid progress made in genetic research. This has involved a massive understanding of the 'how' and 'what' we are as natural entities; from genomic mapping to genetic manipulation. Understanding these fundamental natural laws will progress our fight against disease. The research gives us power to genetically modify species, a kind of new evolution. This ability to affect the characteristics of a species will lead to revolutions in medicine and food production. It also means we can affect the characteristics of all species including ourselves. This power needs to be very carefully discussed and monitored because it positions the human relationship to the natural environment in a new way. Any abuse or misguided use of this technology could threaten our own survival. This is why bio-ethics is now a hugely important area. The double whammy of climate change and genetic engineering put us at a pivotal place in the evolution of our relationship to Nature. We have a kind of ownership and, like much ownership, it implies responsibility.

All this feels like the reasoning of the mechanical, the machine view of our cosmos. Much was, and is, made of Darwin's theory being

in opposition to the immanent view of the cosmos; that in some way the evidence presented is antithetical to our need for spirituality and imaginative play.

Certainly it did give a view of the natural world that contradicted many religious texts. However, I think that the two models can be better positioned not to oppose but to include each other.

A mind in the soil

The idea of a cosmic principle or mind that is both immanent and transcendent in our relationship with Nature has been under critical review over the last century. The psychological, philosophical, theological and ontological complexity of God as a 'mind' within the cosmos in direct opposition to a cosmos seen as a 'mindless' machine applying the natural laws has coloured our understanding of Nature. It seems to me that when we look at the natural world it is difficult not to conjecture the existence of a mind in the cosmos. At times of difficulty, or thankfulness, the transcendent entity that is always beyond us is called on.

In my garden the evidence to my senses is compelling in the inference of a transcendent condition or state of being that surpasses physical existence; an invisible force that seems beyond and independent of existence whilst pervading all living things, immanent. In many ways this could be a description of the invisible forces that drive the natural laws, Nature.

The problem may be that my rational self says such an entity is implausible whilst my subjective self welcomes an entity that is always beyond the immediacy of events. I know that Nature is independent of human existence and purpose.

The existence of a mind within the invisible forces of Nature does allow me to bring human nature within Nature and to see that human purpose is part of these invisible forces. This may be a delusion, but it does assist the human species to feel that they have a stake in the cosmos. In some ways I see the laws of Nature as a mind in and of itself; and the transcedent, immanent entity as a necessary human construction formed to connect us to the natural laws. For me this is an abstract construction, true for all places and times and therefore not subject to specific contexts.

I am sad that this construction has been adopted, in varying forms, by religions; spirituality with names. The very act of naming has contextualised the abstract construction. It is within this contextualisation that many of the religious have deployed the particular, often textual, arguments that devalue the construction. Religion has argued and killed in the name of a mind that we cannot yet comprehend. Religion has also created some of humankinds most sublime rituals of faith.

We are young as a species. In the manner that a child conjures up invisible friends as entities for comfort and company, perhaps we need similar constructions to help assuage our loneliness amidst Nature; a kind of company and comfort that helps us accommodate the wonder and mysteries of the cosmos.

In a society driven by audit, the notions of transcendance and immanence sit uncomfortably. I think that belief, importantly subject to contextual change, is different to faith which is an absolute trust and not subject to context. Faith is more a human law.

In the debate between 'machine' and 'mind' views of our cosmos, religions have become borders that seem unnecessary to me. In my garden I view the cosmos as both mind and machine. I see it as an unknown fusion of the subjective and objective held within a relationship that is beyond my comprehension.

Responsibility in and of the soil

For many generations we have endeavoured to communicate with Nature's phenomena, with sky, seas, rocks, plants and animals to try to find a sense of ourselves. We have reached a point where we can see the mechanical aspects of that order, to see ourselves in the machine. Indeed, we have the knowledge to interfere in natural selection, our own and others. We can alter evolution. Much of this has been rapidly acquired knowledge of our natural biological mechanisms.

Earlier in this text I referred to natural science and human science, with one being of natural laws and the other of human contexts. What seems evident is that natural law and human context are being refigured. Our ability to affect genetic determinants moves the two sciences into an overlapping relationship. Natural science and human science, laws and contexts, become a relationship unique in our history. As humans we have affected our natural environment; we could now begin to review the relationship of the human and natural sciences. As a species we seem to be moving towards a situation where natural selection is being supplemented by human selection; the anthropogenic within the natural.

Human selection, when distorted by eugenics, has resulted in cruelty, degradation and inhumanity. The assumption that human genetics should dominate social context and conditions has had consequences that all humanity condemned. The full title of Darwin's seminal work is 'On the Origin of Species by Means of Natural Selection, or the Preservation of Favoured Races in the Struggle for Life'. The expression 'favoured races' is one that has implications for all of our species. Darwin's use of the phrase has to be historically located, clearly it was not written as an announcement of vile experiments to follow, nevertheless, the expression does foreshadow a warning.

As a species, it is now imperative not to be divided by racial borders where human selection favours only some. I believe if that happens then Darwin's final phrase 'in the struggle for life' will be disastrous for humanity. I am not suggesting that current biological research is eugenic, rather that it is crucial to better understand human relationships to each other, to other species and to the natural world. In particular, I would suggest that the overlap between human and natural sciences needs to emerge from a moral and ethical understanding of how we work with our scientific developments and applications.

It is very apparent that most of us may have some knowledge of the effects of this scientific revolution but few fully understand the language, meaning and implications of the science. To better understand we need better communication. This is not simply a material revolution but also a philosophical one. Ethical understanding is not the exclusive understanding of the wealthy and powerful nations. In a truly global society we need to see a moral and ethical context across all peoples. This may sound fanciful given all the conflict in the world, but given that Nature does not comply to any human border and is independent of our existence, then there is a need for a human endeavour that crosses all borders; a kind of human nature that understands human purpose and its connection to existence.

Rapid scientific development requires great responsibility. Part of that responsibility will be formed by our subjective wonder at the mysteries we observe in our cosmos. We can understand the empirics of biological advances whilst still loving those things that are evident to our senses.

I do believe if we can balance these twin progressions then we will live in a better and more tolerant world. We need to wonder at the cosmos and to speculate on all its mysteries. As a machine, it has to be experienced as a machine that is like no other. A machine that has no edges and that may be able to evolve new natural laws. A machine that can have the subjective melded into the objective. From my garden, I experience such an entity. In the garden, my purpose is to better understand it. The soil we construct needs to be composed of this knowledge.

Drawing Threads

Being less confused

This essay has involved reflective reviews at several points in its development. At each stage of the reflection, a catching up so to speak, the next set of thoughts and questions begins to emerge. In this manner the text is alive to new possibilities, meaning that I am never sure where it will end. In many ways it is simply a continuum of what I do and how I work. I have tried to look at how our human nature resides within and without the natural. I have done this through observations and actions that are common in our experience of working and being in a garden. I have considered our garden as a clearing in which entities independent of us can be revealed and our connections to them made clearer. It has been a set of thoughts that attempt to place us somewhere.

Early in the process I expressed my confusion about our place in all of this complexity. I have tried to construct a partial cartography that reveals some of the routes and territories that we need to understand so we can see and experience the clearing. Like all maps the clearing is evolving and changing. It is not the territory that we experience through events and actions. My hope is that, by trying to tease out constituent elements that inform our relationship to Nature, I stimulate further consideration of our connectedness with the natural. For myself, I do feel less confused and more able to apply what I have learned to the questions that humanity will be asking in this Anthropocene era.

When I see a squirrel digging up the back garden, or a fox sleeping in the hedge, or a plant thriving after some simple intervention, or the bees working on the lavender, or the bats flitting around at dusk, or the unfolding of a flower, as complex in its geometry as it is stunning in its colour, I am conscious of my relationship to these events.

I do think of affect and effect, of machine and immanence, of natural and human selection, of genetic and social predisposition, of human and natural science, of existence and purpose, of the contingent and the contiguous, of borders and openness, of thresholds and clearings, even of what a pet shop may imply, but above all I wonder at it, and care. I care because I wish not to be lonely in the cosmos.

John Newling

Ecology Works 2004–2011

Chatham Vines

Ark Lab: The Noah Project

The Noah Laboratory (Constructing Soil)

The Lemon Tree

The Clearing

Root Zone

Local History

688 Days: Remembering the Lemon Tree

Natural History

Synthia II (code / soil / life)

Chatham Vines

2004–2006

In *Chatham Vines* the artist created a living sculpture by planting thirty-two Pinot Noir grape vines in the central aisle of St John's Church in Chatham, Kent. Using hydroponics, a method of growing plants in a minutely-controlled environment using nutrient-rich gel instead of soil, the vines were grown for a period of one year. At night the church was lit from within and two cameras continuously streamed images of the project to a dedicated website, enabling the artist and his team to carefully observe the vines progress, whilst also reconnecting the church to an outside world that had long forgotten it.

Whilst the vineyard successfully produced grapes, the plants clearly suffered from 'stress' due to being subjected to intensive growing conditions, 24-hour light and, at one point, an experiment to simulate spring in mid-winter. In spite of this, the vines produced several bottles of wine, some of which was consecrated and used for communion during the Dawn Eucharist at Rochester Cathedral on Easter Day in 2006. The transformation of the grapes into wine acted as a poignant metaphor for change in Chatham and signified the new life that the project had brought into both the church and the community around it.

Chatham Vines Wine, 2006
Wine produced from the hydroponic vineyard

Chatham Vines, 2004–2006
Installation of the hydroponic vineyard in St John's Church, Chatham, UK

Ark Lab: The Noah Project
2008

'Newling made his call for people to live in a 'biocity' publicly in *Ark Lab: The Noah Project*, a series of performances that formed the preliminary stage of *The Noah Laboratory*. On 18 February 2008 he stood outside Primark clothing store on Lincoln's High Street and gave away free Playmobil sets of Noah's Ark toys to passers-by. A day later, on 25 February, he stood at the level crossing by Lincoln's railway station and gave out border spades, and then finally on 13 March he stood outside the tourist information centre on the High Street and gave out birdhouses. Taken individually these interactions with the public might seem quirky and innocuous, but when they are witnessed through the documentary photographs he included in the project's free newspaper, the underlying narrative of these 'gift-giving' events becomes explicit.

The three photographs in the newspaper that illustrate 'Gift I', the Noah's Ark, are set in a completely urban landscape. Only in the last is a tree included, glimpsed half hidden behind a telephone box. These are images of a world on the point of environmental destruction, the Noah's Ark toy a symbol of the biblical flood that wiped clean the surface of the world. In the images for 'Gift 2', the border spade acts as a gardening tool, giving the opportunity to make a new start, not on an agricultural scale but on a more human scale, in a garden. Behind the recipients of the gifts, trees can be seen now, a tangle of leafless branches reaching into the sky, nature asserting its presence, caught in the space between death and promised life. In the photographs of the final gift, the birdhouse, the recipients stand between two leafy bushes. In the largest photograph a man holds both an enormous pet rabbit and the birdhouse. Here is an image of the biocity in which the oppositional tension between city and forest has been broken down; allowing a space for birds to nest and the natural world to breathe.'

Richard Davey, from his book *Spinning – Nature, Culture and the Spiritual in the Work of John Newling*, 2013

If Noah was a Pirate

At a time when many ideologies have all but blown themselves out; when meaning seems to have become meaningless in a strangely nihilistic void; when, in our sophistication, we find as our ancestors did that we are fearful of the unpredictability of Nature; when knowledge as causality seems to become increasingly predetermined; it seems heaven sent that there remains the oasis of artist's expressions.

Artists are pirates. We plunder the borders of all known subjects seeking to express our subjective selves and, in so doing, loosening the very apron strings of institutional rationalism. We need these works of expressive uncertainty to remind us that the nihilistic void can be challenged and knowledge is not entirely predetermined.

Less of a subject and more of an essential fluid, art has a value that can never be truly determined, asking, as it does, to trust in its risks. If Noah was a pirate he would raise his flag of smiling death, and start loading his boat with works from artists, not entirely in remembrance, but more to assist in the propagation of a future evolution.

John Newling, 2008

The Noah Laboratory (Constructing Soil)
2009

Newling's residency at the The Collection in Lincoln, between January to April 2008, resulted in a major project, the first in which the artist began experimenting with the production of soil. The work consisted of three large metal cages positioned in the centre of the gallery. The first was a collection and distribution centre for a newspaper containing his essay, *From My Garden* (republished within this book on pp 15–29), which was distributed around the city and on the train between Nottingham and Lincoln. The second cage was dedicated to recycling, where papers that were left unused around the city were collected and then shredded. A third cage contained a hand-operated rotating compost tumbler. The artist added the shredded newspaper to the tumbler, vigorously turning it every day. The drums eventually converted the newspapers into a compost capable of sustaining life.

The Noah Laboratory was the first instance in which Newling converted provocative and profound texts into soil as a symbolic act and was the first work in which an 'ecosystem' was embedded into a project. This approach was carried through to many of Newling's works thereafter and is reflected in many projects within this book. The soil from *The Noah Laboratory* was subsequently used by the artist to grow a lemon tree, which is documented in the book *The Lemon Tree and Me* (p 38).

The Noah Laboratory (Constructing Soil), 2009
Aluminium cages, artist's newspapers, compost tumblers

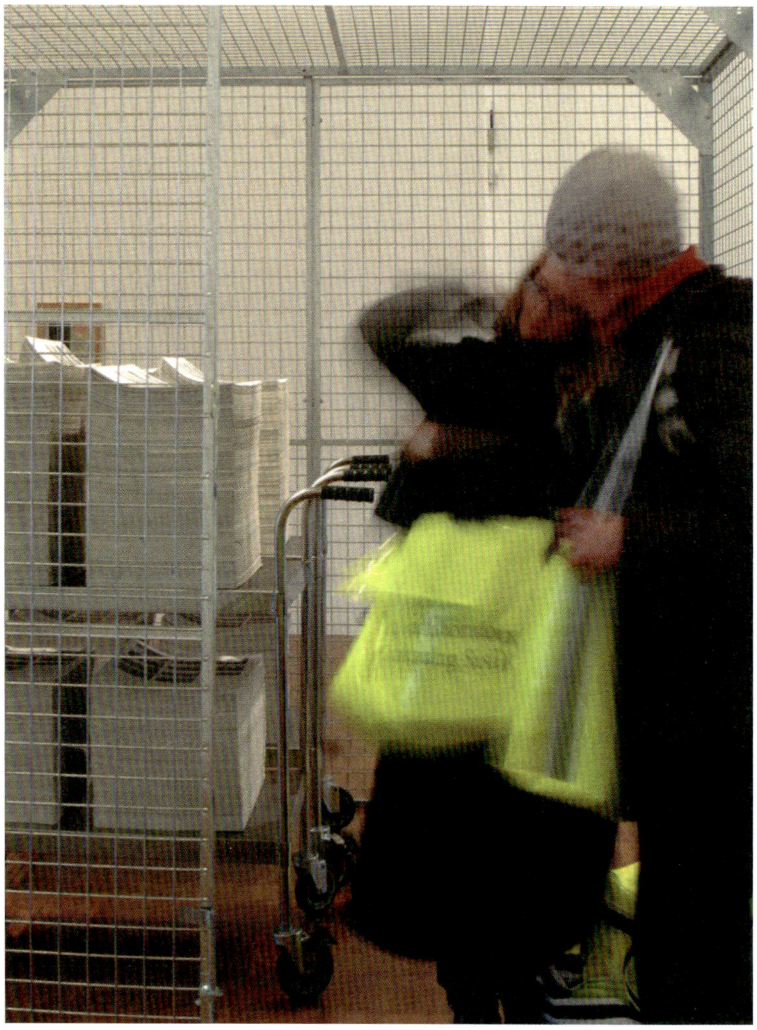

John Newling processing shredded texts into soil using a domestic compost tumbler.

The Lemon Tree
2009–2015

The Lemon Tree & Me is a diaristic account of an intense period between March 2009 and February 2011 in which the artist nurtured and carefully observed the growing of a lemon tree using soil produced as part of *The Noah Laboratory* at the The Collection in Lincoln. Once grown, Newling returned the lemon tree to the gallery to be maintained in a hydroponic grow tent. Whilst the artist was able to maintain the tree on a form of 'life support' for a short period of time, it eventually perished. The project captures the poignant relationship between humankind and nature, one of symbiosis and of responsibility, and highlights the fragility of our ecosystem in the human-centred era of the Anthropocene.

The artist went on to strip the roots bare revealing a complex network of the plants once active and thriving growth system, creating the work *688 Days: Remembering the Lemon Tree* (p 46). The root was integrated into a later work that was shown as part of *Ecologies of Value* at Nottingham Contemporary in 2013 (p62).

Whilst *The Lemon Tree & Me* takes the form of a book, the artist considers the act of documenting caring for the lemon tree as an artwork in its own right. Newling has maintained a practice that has always developed texts alongside artworks, as both a critique, exploration and complement to his work. *The Lemon Tree and Me* is followed by two more books that are part of the artist's 'ecology works', *Nymans Language* (2016) (p 100) and *Dear Nature* (2018) (p 130).

The Lemon Tree and Me, 2015
Artist's book

The Lemon Tree, 2009–2011
Lemon tree, hydroponic grow tent

The Clearing

2009

The Clearing was a series of artworks commissioned by curator Jennie Syson along the urban hinterland of the banks of the River Trent in Newling's hometown of Nottingham.

The artist worked in a laboratory called BioCity, the largest bioscience incubator of its kind in the UK, just a short distance from the Trent. During his time there, working closely with BioCity scientists, the artist grew six beech trees under hydroponic lights, monitoring their development, feeding the plants solely with liquid soil.

During his time at the laboratory Newling collected and copied key documents relating to the history of the spaces adjacent to the River Trent and the industrial wastelands of Nottingham, which emphasised the shifting societal values and changes of land use in the area. Following a similar process to *The Noah Laboratory*, the artist shredded the documents and created soil from the decomposed text in his garden at home. The artist then removed the beech trees from their artificial life support system and replanted them on the riverside, in a community area in Beeston, Nottingham, using the 'symbolic soil' he'd made to bed the trees down. The nine beech trees were planted in a circle, with the intention of creating a natural frame, stage or venue for artists, where ideas could be shared; a new kind of laboratory within the community, created from trees, planted within a terroir of local history.

Newling has frequently worked with experts in their field in an attempt to gain a deeper understanding of a particular type of knowledge or specialism. The very act of working in an environment where the business activities may be at odds with natural systems shows the artist's curiosity and admiration for science, but could also be seen as a subversive act.

The Clearing, (first phase), 2009
Birch trees growing in hydroponic conditions at BioCity laboratories, Nottingham

The Clearing, (second phase), 2009
Birch trees in the artist's garden being prepared for outdoor planting

The Clearing, (first phase), 2009
Growing experiments, at BioCity laboratories, Nottingham

The Clearing, (second phase), 2009
Permanent installation of the birch trees, Beeston, Nottingham. The two images show several years of growth.

Root Zone

2009

In *Root Zone* the artist cultivated the small area of land under the tree on the street outside his house. Once the soil had been prepared, Newling planted numerous herbs and strawberry plants in the 'zone'. Local people continue to use the plants, harvesting the herbs and taking cuttings from which to grow their own.

Root Zone continues my work on the relationship between the natural environment and me; a relationship of meaning and materiality that constructs, cultivates and reviews a poetics of responsibility. It is a relationship that advocates an intelligent ecology based on values that are immanent in the complex workings of nature. Between ethics, ecology and aesthetics I hope for my work to find a new ground in a generative programme of intensive care to influence our responsibilities as gardeners of the public domain.

In recent years great attention has been paid to the complex dynamics of ecological systems and how, in practice, we can think in terms of urban, social, political, and even cultural ecologies. My soil works are situated on this shared ground of possibility, where the artificial distinction between nature and culture will no longer make sense. Seeing Nature as both affect and effect, tied into the multiple ecologies of humanity, we can re-evaluate its transformative potential in the complex workings of everyday life. A social ecology, in a state of co-creation with the natural environment, could provide the practical grounds for addressing the question of how to properly constitute the well-being of contemporary life.

Local History

2010

I have been collecting and pressing many leaves and flowers from the *Root Zone* site. These pressings will then be placed into all the books in the local history section of the library in Sherwood. I will be placing one or two of the pressings into each book. I have, on a few occasions, been reading a book and come across a pressed flower or leaf and know that it is a strange and oddly personal encounter.

From early July people taking out local history books will come across these delicate, resonant markers. I am curious if the pressed leaves will be removed or left in place.

Root Zone, 2009
A selection of herbs planted around the base of tree

Local History, 2010
Pressed plants and flowers, local history books

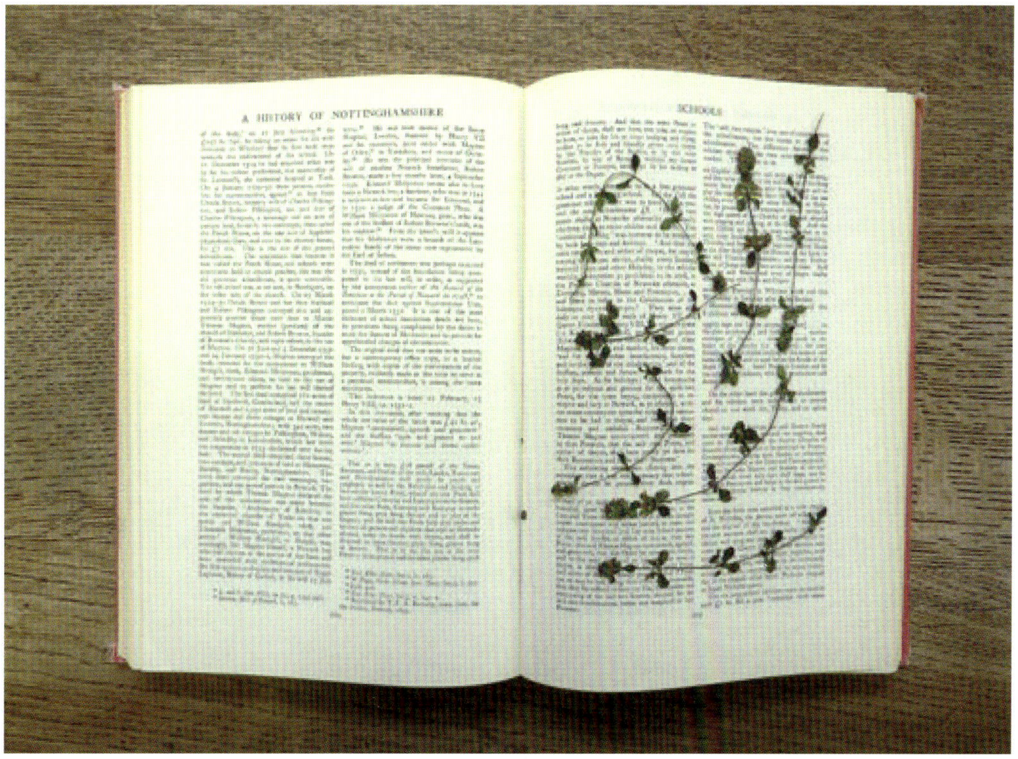
A HISTORY OF NOTTINGHAMSHIRE
SCHOOLS

688 Days: Remembering The Lemon Tree
2011

688 Days: Remembering
The Lemon Tree, 2011
Lemon tree root

Monday 17th January 2011

Today I received an email from Jeremy Webster of the Lincoln Collection informing me of the sad news that, despite every effort made, the lemon tree was dead. It was the kind of news that has a visceral affect and, whilst it was remarkable that the tree lasted for as long as it did, it was still unexpected.

Back on Tuesday 24th March 2010, when I received the somewhat bedraggled lemon tree from Jonathan Casciani I had little notion of the quantity and quality of experience and knowledge that *The Lemon Tree & Me* project would bring. Now, 95 weeks later, *The Lemon Tree & Me* is coming to an end. The lemon tree was with The Collection for 308 days, which was a great achievement on their behalf.

It is the closeness of the relationship between the lemon tree and me that has stayed with me and continues to engage my thoughts. The project sought to privilege an intimacy of knowledge gained through observation and experience. Such closeness tried to overcome the material, social and ethical distance that can occur in works that attempt to comment on situations or conditions that are evident but not experienced.

The relationship had no formal specifications of how to represent it but, nevertheless, seemed at times to open and connect thoughts and experiences inside a domain of knowledge that tried to be exhaustive in its organisation. The organisation of the relationship seemed to be governed through many forms of knowledge; experiential, procedural, social, conceptual, qualitative, experimental, descriptive, quantitative and self-knowledge.

Imbedded in the relationship, there was a sense of a transcendental experience; an early knowledge remembered. This acted as glue within the co-relationship, underscoring, connecting and driving slow time and the need to respond intuitively. It enhanced the possibilities of imagined spaces between ecology and aesthetics.

There were times when the intense relationship felt cut off, isolated from everyday experiences on an island of conceptual and material localism. Conversely the island experience did teach me much about 'what we have' as I tried to make wider connections from within the relationship; bridges from the island.

The project was drenched in intricate interactions connecting within and without the complexity of the relationship. As such, negotiating such complexities embodied the realities of uncertainty and trust as essential constituents of these bridges.

The values of *The Lemon Tree & Me* were grounded in a wonder of our relationship within the natural environment and sought to transpose this sense to other situations that are common to us all.

Tuesday 8th February 2011

Jonathan brought the lemon tree back from the Lincoln Collection.

Wednesday 9th February 2011

Today I removed the lemon tree from the soil. I washed the root system and preserved it.

This last vestige of the tree sits in my studio like some ancient relic. But what an amazing relic it is. The roots of the lemon were the core of where the constructed soil met the natural systems of the tree; both ecological station and vehicle; a matrix as both heart and an echo of the complexities of the relationship.

The Lemon Tree & Me *evolved as an intense co-relationship where the material expression of the work was the result of that relationship; a working from life; a love song; sung for 688 days.* The Lemon Tree & Me (*Tuesday 24th March 2009–Wednesday 9th February 2011*).

Natural History
2010

Natural history is a series of small sculptures that were formed during the *Peterborough Soil* installation (January–February, 2010). Each of nine small spheres was taken out of the compost tumbler over a period of eight weeks, one per week. The series of nine spheres reflect the transition from paper text to soil. These are the first of a number of sculptural works that form from moulding the soils the artist produces within his circular compost tumblers. The work is one of the most explicit examples of the way in which Newling's work emerges out of an analytical, experimental process that at times emulates the work of scientists.

Natural History, 2010
Shredded paper text, soil

Synthia II (code / soil / life)

2010

Synthia II (code/soil/life) consists of a soil constructed by shredding the written code for Craig Venter's synthetic life form 'Synthia' and blending it with vegetable matter grown in the *Root Zone* (p 44) land outside the artist's house. The soil was constructed over an 18 day period that deployed a closeness of observation and experimentation in order to convert 634 pages of code and text into a material that could sustain life. The soil was constructed in the hope that it would enable the growth of a Moringa Oleifera 'miracle tree'.

Venter's 'Synthia' seems at odds with a social ecology, in a state of co-creation with the natural environment. The attempt to grow a Moringa Oleifera tree in a soil largely constructed from Venter's code is a renewing of the vital need for a co-creation between us and the natural environment.

Synthia II (code / soil / life), 2010
Shredded paper code, soil

A growing experiment to identify if the soil was capable of sustaining life (bottom right)

Ecology Works 2011–2014

Miracle Trees
Transfer: The Relic of a Miracle
Lungs
Miracle Maps
Common Sense (Thomas Paine's Soil)
Ecologies of Value
Value; Coin, Note and Eclipse
Blanket
Miracle Trees
From a Garden of Walking Sticks
Between Breathing and Spinning
Fall
Green Through Gold, Gold through Green
Small Sculptures (Ecologies of Value)
Generosity Plates
Human Nature Tables

Miracle Trees

Nottingham Contemporary, 2011

In *Miracle Trees* Newling germinated and attempted to grow eleven Moringa Oleifera trees. Often referred to as the 'miracle tree', these 'generous' trees are native to the southern foothills of the Himalayas in northwestern India. Gram for gram, Moringa leaves contain seven times the vitamin C found in oranges, four times the calcium and twice the protein found in milk, four times the vitamin A in carrots, and three times the potassium found in bananas. It is for this, and other extraordinary properties of this tree, that it has been referred to as the world's most generous tree.

As part of his research Newling closely observed the growth of the trees, making pressings of the Moringa leaves and complete trees. Some of the leaves from the Moringa were pressed in selected books in Nottingham Contemporary's study, complemented by a video documenting Newling's process of making the soil, caring for the plants and pressing their leaves.

Hydroponic grow box containing Moringa Oleifera at Nottingham Contemporary

Miracle Trees, 2011
Nine pressed Moringa Oleifera mounted onto board
Exhibited at Wellcome Collection, London, 2014

146 Days

[illegible]

Pressed Moringa oleifera trees

9 framed works

Transfer: The Relic of a Miracle

2011

On Wednesday the 9th of March I planted some 18 Moringa Oleifera 'miracle tree' seeds in small pots and placed them behind the curtain in the study at Nottingham Contemporary. On Friday the 18th of March the first shoots of the germinated trees were seen. Of the 18 seeds planted, fourteen germinated. It was from one of the 4 failed seeds that the 'miracle tree' tattoo emerged.

On Monday 28th of March I brought the plants back to my studio. I checked the soil of the seeds that had not germinated. To my surprise I found one seed that seemed to be growing upside-down in the soil. I carefully removed the seed and re-planted it. Whilst the tree did grow it stopped growing after a few days. I rechecked the seed and found that it had no root production. Despite the seed clearly being damaged it was amazing to think that it still managed some growth and leaf production. The power of the evolutionary imperative to grow is staggering.

The shape, provenance and ontology of this tiny tree seemed to lend itself to possible material transformation. I decided to preserve the small tree. Some five weeks later the tree was pressed. It measured 4cm in height by 1.5cm in width.

It is the closeness of the relationship between the miracle trees and me that has stayed with me and continues to engage my thoughts. The work, like many of my projects, seeks to privilege an intimacy of knowledge gained through observation and experience. Such closeness tries to overcome the material, social and ethical distance that can occur in works that attempt to comment on situations or conditions that are evident but not experienced.

The tiny tree reminded me of the kind of desert island that is beloved by animators. It did seem to echo the occasional sense I get of being isolated on an island of conceptual and material localism within the relationship. That said, this isolation, even loneliness, tends to be short lived as the experiences and subsequent knowledge of the relationship makes connections in and through the complexity of other things.

It was with the combination of the 'island' metaphor and the notion of 'closeness' that the idea of a transfer tattoo emerged. Close here becomes skin with an image lying between our corporeal and experiential senses; a place where art often hides out. Transferring is a common concept in the practice of art. Whether it is the complex exchange of meanings constructed through moving something from one place to another or, equally complex, the transference of material and symbolic meanings within the work, art restlessly transfers our material and psychological selves.

The work is the transference of the Moringa Oleifera tree from its organic form, to a reproducible material of adornment. The tattoo transfer becomes an expression of this transference; an image guised as a desert island but, behind the guise, a relic of a miracle. To wear it is to briefly acknowledge the nurture that is Nature from an island that is a miracle tree.

Moringa Oleifera seedling

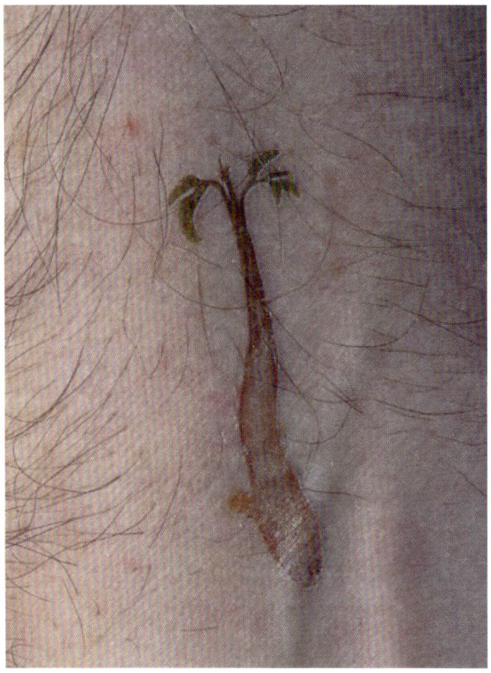

Transfer: The Relic of a Miracle, 2011
Temporary tattoo transfer

Lungs
2012

In a derelict sanatorium in Brehmena, Poland, Newling fashioned a pair of 'lungs' from Jersey Kale cabbage leaves he had grown in his garden, held together with Elastoplasts with a central throat-like twig. The work is witty, tragic and philosophical. Its 'Heath Robinson' nature comments on humanity's short-term attitude towards our very existence on earth, while simultaneously offering a gesture of healing. By constructing the work in front of a live audience the artist breathed life back into the desolate space that is the sanatorium, that stands as a symbol of man's current predicament on earth.

Lungs, 2012
Pressed Jersey Kale, Elastoplast, stick

Lungs (performance), 2012
The artist making *Lungs*,
Brehmena, Poland

Miracle Maps

2009–2011

Newling constructed this series of 3 works using many pressed fragments of Moringa Oliefera 'miracle trees' that the artist had grown in his studio at home and as part of the installation at Nottingham Contemporary.

The works are a contemplative act of constructing an imagined cartography, which went on to form several major projects, in particular *The Map Room of the Last Islands* (2015) (p 90).

Mapping

I have made many works that have the word 'map' in their title. Throughout my practice I have referred to the cartography of things; a continual mapping of thoughts, concepts and materials. In parallel with this I have also referred to the phrase 'The map is not the territory'; a phrase commonly used by the Situationist group in the 1960s. It could be argued that a desire to map – twinned with an understanding that maps are not the actuality of our experience of place – could be seen as oppositional. For me, the cartographies I often refer to are tools that I use to map the complex interconnectivity of thoughts, materials and places. These are not precise maps but ones where the borders of things slip and change constantly and relationships between us and our environment are fluid.

Maps are often associated with destinations. An abiding approach in my work is to gather sets of processes, materials, thoughts and places to see where they may begin to take me. This is not unlike the sense you get when walking without a destination in mind. Whilst there is often a conceptual or geological premise to the work it is allowed to meander through processes and material changes to open up my sense of its cartography. It is only in the middle of these works that a destination / artwork emerges. In this manner the map becomes both territory and conceptual, material destination. I guess much of my work is trying to find where 'here' is. In this I know the maps become guides in this search, each work adding to my sense of place, where place refers to the complexity of the ecology of us within all the co-relationships we have with other things and species.

Miracle Maps, 2009–2011
Pressed and constructed Moringa Oleifera trees

1
2
3
4
5
6
7
8
9
10
A
B
C
D
E
F
G
H
I
J
K
L
M
N

1
2
3
4
5
6
7
8
9
10
A
B
C
D
E
F
G
H
I
J
K
L
M
N

1
2
3
4
5
6
7
8
9
10
A
B
C
D
E
F
G
H
I
J
K
L
M
N

Common Sense (Thomas Paine's Soil)

2013

Newling was asked by Robert Blackson, curator of the Temple Gallery in Philadelphia, to make a soil project. After some research he found out that Thomas Paine, after a meeting in London with Benjamin Franklin, emigrated to Philadelphia where he worked and published many pamphlets that were both visionary and revolutionary. Newling decided to construct soil from copies of Paine's 'Rights of Man' and 'Common Sense' texts. Published in 1791 this book, along with other Thomas Paine texts, are acknowledged as being revolutionary in spirit.

Over a period of 31 days Newling constructed a soil/compost made from 9 complete copies of this revolutionary book. The soil consisted of approximately 75% text with the remaining 25% being made up of vegetative matter as a catalyst for the transformation from books to soil. The process produced some 7.5 cubic feet of soil that weighed approximately 30 kilos.

Finally, the soil was rolled into balls and dried, then, the 30 balls were stacked to form a structure reminiscent of a cannonball mound; both a response to Paine's revolutionary thought and as cannonballs that sustain rather than take life. Sadly, the American authorities would not allow the soil entry to their country. The work became one of several soil works as part of the exhibition *Ecologies of Value* at Nottingham Contemporary in January 2013.

Common Sense
(Thomas Paine's Soil), 2013
Paper, soil, 6 copies of 'Rights of Man' by Thomas Paine

Paine Rights of Man, Common Sense and Other Political Writings OXFORD
Paine Rights of Man, Common Sense and Other Political Writings OXFORD
Paine Rights of Man, Common Sense and Other Political Writings OXFORD

Ecologies of Value

Nottingham Contemporary, 2013

The 2013 exhibition *Ecologies of Value* was the seminal exhibition that explored the natural world and the social and economic systems that govern society, such as money or religion. These are themes that have consistently occurred throughout Newling's practice since it began in the mid-1970s. The show linked the plant-based 'ecology works', that have become a signature of the artist's work since the early 2000s, and early works that took many different conceptual and sculptural forms.

For *Ecologies of Value* Newling grew and harvested over eighty Walking Stick Cabbages (Brassica Oleracea), which were used as the key ingredient to the new works that he created for the show.

The artist in his garden growing Jersey Kale or Walking Stick Cabbages in 2012 in preparation for *Ecologies of Value*, 2013

Value; Coin, Note and Eclipse

2012

To make *Value; Coin, Note and Eclipse* the artist grew more than 80 pots of Jersey Kale (Brassica Oleracea) in his Nottingham garden. Newling harvested the leaves of these plants at different stages of their growth, cutting and drying them to use as materials. Applying gold leaf to their surface, and transforming them into a kind of currency, he also used these leaves to create replicas of £20 notes, and the pattern of an eclipse.

The result is a series of two-dimensional works exploring time, currency and value, reflecting on the evolution of money as a form of transaction that appears to bring a divide between us and our environment. The work deploys changing eclipses between gold and green as an evolving debate between the natural occurrences of the earth and our pursuit to acquire monetary wealth. It reads as a long 'conversation' wherein our coexistence with nature seems to be slowly eclipsed as coin and paper money become dominant values.

Value; Coin, Note and Eclipse, 2012
27 framed, pressed, gilded and reconstructed cabbage leaves

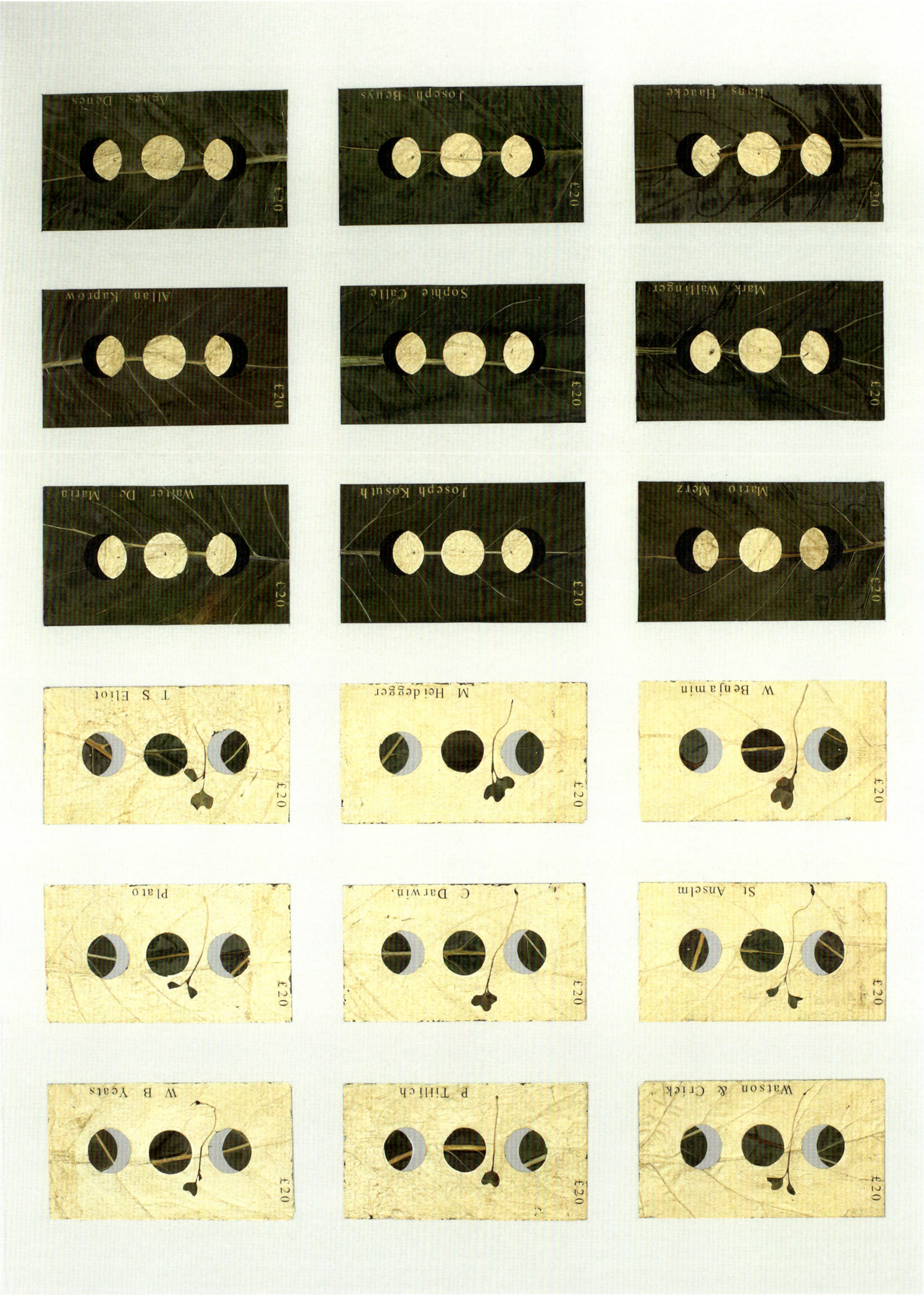
Agnes Denes
£20
Joseph Beuys
£20
Hans Haacke
£20
Allan Kaprow
£20
Sophie Calle
£20
Mark Wallinger
£20
Walter De Maria
£20
Joseph Kosuth
£20
Mario Merz
£20
T S Eliot
£20
M Heidegger
£20
W Benjamin
£20
Plato
£20
C Darwin.
£20
St Anselm
£20
W B Yeats
£20
P Tillich
£20
Watson & Crick
£20

Blanket
2012

Blanket is constructed from Jersey Kale cabbage leaves the artist grew in his garden and was made for the *Ecologies of Value* exhibition (p 62) at Nottingham Contemporary. The leaves were pressed, layered and preserved to form a 'blanket' that was then framed. The blanket is a domestic object that comforts every human throughout their lives from birth to death. Newling uses the blanket as a metaphor for the need for the human species to re-establish an essential co-relationship with nature.

Blanket, 2012
Pressed Jersey Kale Leaves

Miracle trees

2012

Two silver-lined hydroponic growth tents contain young Moringa Oleifera 'miracle trees', a continuation of an earlier project at Nottingham Contemporary in the gallery's study (p 52). Through carefully regulated levels of light, carbon dioxide and water, the tents simulate the trees' native climate in the Himalayan foothills of north-western India. Newling became interested in the plant when researching the species' uniquely generous properties for healing and nutrition.

Ecologies of Value, 2013
Installation at Nottingham Contemporary

From a Garden of Walking Sticks

2011–2012

In his work *From a Garden of Walking Sticks* Newling grew over eighty Walking Stick Cabbages (Brassica Oleracea) and installed them in the gallery. The cabbages changed physically over the course of the exhibition as they dried and hardened. Newling became interested in the uses for Walking Stick Cabbages – they have the potential to grow to enormous heights (up to 5 metres) and were traditionally cultivated as cattle feed. Once their leaves have been eaten, their trunks can be stripped and dried to produce very strong, yet very light, walking sticks.

Ecologies of Value, 2013
Installation at Nottingham Contemporary
(left) and the Newlyn Art Gallery (right)

The act of gardening attempts to restore a set of values to materiality. In gardening we live within the authority of nature and experience its immanent vitality in the emerging complexity of Life. *From a Garden of Walking Sticks* imagines a space for a creative dialogue between ecology and aesthetics; germinated from the seeds of practical experience, the artwork evolved as my relationship with the plants matured.

The final phase of this project sees the walking sticks cut, sanded and varnished ready for use.

From the inception of this project I knew that, if I was successful in growing the Jersey Kale, I would arrive at a moment when others could take ownership of the walking sticks. Not unlike seeds being subject to the wind, the walking sticks become materials of exchange, transformation and value dispersed through others.

I see *From a Garden of Walking Sticks* as a poetic monument for an emerging social ecology where care is seen through the support of each walking stick. In future years I will wonder where the sticks are now and how their material history has evolved through touch and place.

From a garden of Walking Sticks, 2012
Dried stalks of Jersey Kale, rubber feet
Installation at Ikon Gallery 2020

Between Breathing and Spinning

2012

In 2012 Newling was invited to make an installation-performance for *Contexts 2012* – The 2nd Sokolowsko Festival of Ephemeral Art, curated by Malgosia Sady and supported by the IN-SITU Foundation. Here the artist created a work called *Lungs* (p 55) which took place in a disused hospital for the treatment of respiratory diseases.

Newling had noticed that the large Jersey Kale leaves he had been collecting and pressing looked like human lungs. For the work at Sokolowsko he performed the construction of lungs and installed the five sets of lungs he'd created in the ruins of the hospital.

In the work *Between Breathing and Spinning* Newling constructed lungs again and placed them onto a lightbox that was used for looking at x-rays. The installation continued vertically, connecting the lungs with two clocks which had a branch construction on their faces that implied a spinning motion. Breathing and spinning have long been visual and material preoccupations in Newling's work.

Between Breathing and Spinning, 2012
Pressed and constructed Jersey Kale cabbages, light box, clocks

Fall
2012

Slow time is important to my welfare. I think all of the growing projects I do force me to be part of a slow time cycle. In this work the clock speaks for itself whilst the leaves as currency begin to look at economies and value. The bundles of cut leaves are exact to the dimensions of a £20 note. They also weigh exactly what £1,000 of £20 notes would weigh. I tried to use the leaves as currency with some success locally but quickly realised that the labour expended over nine months of growing, and the harvesting of the leaves, was of a higher value than its transactional value. Such is the complexity of capital. I do believe that if, as a species, we are to survive we will need to look at models of economies that are possibly a new variant of capitalism or a replacement for the demise of capitalism. I have long been working with transactions between me and nature and believe that such transactions may well inform future economic models.

Fall, 2012
Pressed and constructed Jersey Kale cabbages, clocks, Jersey Kale walking stick, paper banknote wrappers, copper gyre

Green through Gold, Gold through Green

2012

Two clocks. The face of one clock is covered in leaves. The face of the other is covered in gold. Each clock has six evenly spaced circles cut into them. Each circle is the diameter of a 2p coin.

Behind the faces the clocks are running. On each of the minute hands is a 2p coin diameter of gold or leaf respectively. In this manner, every ten minutes an eclipse is seen; gold through leaf and leaf through gold. These are eclipses of value.

Green through Gold, Gold through Green, 2012
Pressed and constructed Jersey Kale cabbages, clocks, gilding

Small Sculptures, Ecologies of Value
2012

This group of small sculptures were produced by the artist for *Ecologies of Value* (2013) exhibition at Nottingham Contemporary. All five sculptures combine elements of earlier works to create a set of new sculptures thus combining earlier lines of inquiry with Newling's thinking at the time of this exhibition.

The first sculpture is a glass cloche containing fragments of pure paracetamol – previously a map of Nottingham made entirely from remoulded paracetamol – sitting on copies of An Essential Disorientation (Newling, 2007). The second is *Common Sense (Thomas Paine's Soil)* (2013) discussed earlier on page 60, the third is the exposed lemon tree root *688 Days: Remembering The Lemon Tree* (2011) (p 46), placed on top of the 'From My Garden' essay in its original newspaper form. The fourth sculpture contains Moringa Oleifera stems with fountain pen nibs, a block of paracetamol containing an impression from a chocolate box, and a four-wheeled church made from cabbage leaves with a copper base, placed on top of a series of book bound cases from his major installation *Stamping Uncertainty* (2004). The fifth sculpture contains a gilded soil ball, a pressed Moringa Oleifera and a newspaper which carries the words 'What do we do to make ourselves feel better?' from the *Make a Piano in Spain* (2009) exhibition at Wellcome Collection, London.

Whilst we can dissect and analyse metaphor, symbolism and concept in all of these pieces, these sculptures represent an aspect of John Newling's practice that is not always recognised. As in nature, Newling considers his work as living entities, exhibitions frequently contain transient dimensions that connect one part of his practice to another and while certain works become 'finished', others remain in flux.

Ecologies of Value, 2013
Installation at Nottingham Contemporary

The Generosity Plates

2012

In 2012 Wellcome Collection became home to a series of Moringa Oleifera artworks, sculptures and interventions. *The Generosity Plates* included nine framed, pressed Moringa Oleifera saplings, a hydroponic grow tent in which grew more saplings for future artworks, a series of bracelets made from scanned pressings of Moringa Oleifera and a series of plates decorated with a photograph of the pressed plant. The plates were used in the Wellcome Collection cafe as an intervention where users of the cafe ate their lunch from plates carrying an image of the world's most nutritious plant.

Wellcome Collection is the art collection and gallery of the Wellcome Trust, a research organisation established in 1936 with the legacies of the pharmaceutical magnate Sir Henry Wellcome. In *The Generosity Plates* the artist presents the world's most nutritious plant against the backdrop of one of the world's most innovative health and pharmaceutical research centres. Through a narrative of growth and decline, with pressed plants held in frames and new shoots being incubated, Newling created an installation that drew attention to the value of natural systems.

We live on a planet that is enormously generous. For many years now I have been investigating the transformations and transactions that weave between people, places and the environment. Many of my works have involved growing and studying a plant or tree. The artwork has broadened and developed my research into plants and the environment we share.

My work involves germinating, growing, observing and learning from the trees. I spend time and resources on them and I have noticed a sense of reciprocation. This generosity is not exclusive to the Moringas, but they have made me conscious of it as an agent of learning and I express that learning through my work.

In July 2012 I made the decision to transform these trees into plates. I had imagined eating a meal and slowly an image being revealed on the plate; an image that is fixed, embedded in the glaze itself, an image of a small delicate tree species. They would be 'Plates of Generosity'.

My hope is that each plate will serve as a reminder of an ecology that is based on the values that are immanent in nature. And that every meal will involve the curiosity that is spinning at the core of the project; a quality, like the food itself, that is fundamental to us all.

The Generosity Plates, 2012
Installation at Wellcome Collection

Human Nature Tables
2012

The *Human Nature Tables* are a material conversation between me and the natural world. In a way they were brought on by my recollection of early school days of tables with natural objects placed on them. Often each object or plant was labelled. I really enjoyed these tables. The *Human Nature Tables* have many objects on them that connect nature to culture in ways that I think about; the relationship between a gilded plaster and a bowl made of soil by way of example.

Human Nature Tables, 2012

Roots, soil, gilded soil, soil covered in coal dust, soil coated in ash

Pen nibs on sticks, soil bowls with pigment, soil bowl without pigment, soil balls, soil balls coated in ash

Soil bowls coated in crushed lavender plants, crows feathers, price label for the crows feathers, soil slabs

Soil bowls, gilded sticks, pressed complete lavender plants, soil slabs with pigment

Ecology Works 2015–2020

The Last Islands

Birches and Berries

Nymans Language

Dear Nature Letter

Be Kind, Be Needed, Be Loved

Eliot's Soil

Trying To Find Where Here Is

Frank's Field

Dear Nature (Book)

Dear Nature (Exhibition)

A Library of Ecological Conversations – Leaves and Me

365 Days and 50-Million-Year-Old Leaves

Burnt Sticks

Dear Nature (Film)

So Sorry

From My Garden

Light and Day

Tillage Works

The Last Islands

2015

'The Map Room of the Last Islands' is made up of sculptures and paintings that refer to a kind of cartography; an atlas of relationships, forms and colours that are the ecology of Newling's practice.

The Last Islands works are made from fragments of Moringa Oleifera trees grown in the artist's studio which form a map-like landscape further defined through paint and gilding. Constructed in a similar way to his earlier *Miracle Maps* (2009–2011), (p 56) the works became a contemplative act of constructing an imagined cartography.

Between 2011 and 2014 Newling installed grow tents in several galleries: Nottingham Contemporary, The Newlyn Exchange Gallery and Wellcome Collection. Each installation generated a source of roots, branches, leaves and stems from the trees which were transformed into the *Last Islands* works.

Island 2, 2014
Moringa Oleifera sticks, acrylic on board, gilding

Last Island – Turning and turning (gyres and geometry), 2014
Moringa Oleifera sticks, gilding, acrylic on board

Map 2, Blackberry Picking,
(2 September 2013)
Thorns, nibs, fruit, blood, gilded Elastoplast

Last Island – Caring for symptoms
(plasters and branches), 2014
Moringa Oleifera sticks, acrylic on board, gilding, Elastoplasts

The Liminal Spinning Tree (Banchory)
Study 1, 2014
Watercolour, pressed Moringa Oleifera

A Tree and Us, 2013–2014
Moringa Oleifera sticks, gilding, watercolour on paper

A Tree and Us (2013) and *Map with Gilded Plasters* (2014), are two works that, although not part of *The Last Islands*, relate closely in style and production method. Both works are constructed from the leaves and stalks of the Moringa Oleifera trees the artist grew during this period and both works continue the artist's interest in new forms of cartography. *A Tree and Us* includes a factory and church motif, the latter occurring throughout many of Newling's works since the 80s. The artwork component reflects the artist's interests and the value systems that are constructed and embedded in different sections of society and their position within nature.

Map with Gilded Plasters, 2014
Watercolour on paper, gilded plasters

Birches and Berries

Attenborough Arts Centre, 2016

Birches and Berries was a living artwork made especially for Newling's show, *Plant Culture* at Attenborough Arts Centre, Leicester in 2016. The work was a small forest of young silver birches, with fruit bushes planted between the trees. The forest and berries were housed in a grow tent containing a hydroponic pod system. Beside it, a stack of 81 bowls made out of differing soils, makes reference to the actions of harvest from within the forest, linking nature to culture.

Like many of Newling's works, the materiality of *Birches and Berries* was ever-changing during the time it lived in the gallery. The berries were harvested over the duration of the installation and made freely available to people in the gallery's cafe. Once the exhibition was over the silver birches and berry plants were moved to an external site nearby where they will be able to flourish.

Birches and Berries, 2016
Soil bowls with pigments

Birches and Berries, 2016
Hydroponic grow tent, birch trees, blackberries

The artist picking blackberries from inside the grow tent

Nymans – A Language From The Garden

2016

In *Nymans – A Language from the Garden* Newling created an alphabet from plant specimens he collected from the gardens at Nymans in West Sussex, creating a language of natural forms.

Nymans was the country retreat for the Messel family who were pioneering horticulturalists that sponsored plant hunters. Newling was invited by the National Trust to create a work in these famous gardens. The artwork invited visitors to follow an 'alphabet trail' of twenty-six sites, taking them on a journey around the garden, exploring the language of its plant species and discovering the origin of each letter. At the place where each sample was collected is a marble block engraved with a Roman letter and its Nymans equivalent. As the seasons change, so too does the visual relationship between plant and letter as leaves flourish and then die.

The conceptual evolution of the Nymans language can be traced to my habit of a daily walk. I walk for many reasons: pleasure, exercise, to get somewhere or to look at things. I often walk to try and settle or focus my thinking. If I have been working on a complex set of ideas my head becomes a bit frazzled and a walk will, sometimes, clarify the question or mode of expression. When I am doing this I have a tendency to drop my head down, looking at the ground as I walk.

I often focus on the pattern of leaves or bits of natural detritus placed across the pavement like symbols on a map. This series of arrangements, constructed through the actions of nature, and changing through the seasons, is a visual language. The arrangements stay in my head and can be seen in my work. The idea of a language from a garden has evolved over many years as I have walked across streets and fields. My languages try to understand and express my sense of place. Over history we have named our landscape to share a sense of place with others. Walking in the garden at Nymans suggested the idea of a language that is specific to its place of origin.

Pages from *Nymans – A Language From the Garden*, 2016

Photographs of pressed leaves and flowers picked by the artist at Nymans garden

Nymans Language, 2016
Etched marble slabs

Nymans Language, 2016
A marble slab installed in Nymans garden

Dear Nature Letter
2016

In 2016 Newling undertook a residency at The Barn in Banchory where he set up another 'soil laboratory' (p 36). During this time he wrote short letters to Nature. Newling produced a series of soil bowls, each marked with a chalked alphabetical character, spelling out one of the letters. The soil bowls were installed in The Barn as a wall-based work, after which they were transferred to a large nearby scots pine forest where they were left as a 'permanent' artwork exposed to the elements.

The letter reads:

> *'dear nature...we need to be needed but forget where we are... sorry...love john'*

Dear Nature Letter, 2016
Pigmented soil bowls

Be Kind, Be Needed, Be Loved

The Barn, Banchory, 2017

Be Kind, Be needed, Be loved was the final work of Newling's residency at The Barn in Banchory, Aberdeenshire. The work was in three separate sections running, in total, for some nine metres in length, each part a simple sign in white text showing each of the three phrases from the work's title.

As part of the first phase of the work Newling stood holding each text in turn at the side of a busy road. For him, the work is a kind of mantra for which, in sharing it, becomes a moment of conversation and reflection.

Be Kind, Be Needed, Be Loved, 2017
Aluminium signs

Be needed

Be loved

Eliot's Soil

2017

As part of the collaborative project *Journeys with 'The Waste Land'*, between Turner Contemporary in Margate and The Herbert Art Gallery and Mead Gallery in Coventry, Newling transformed multiple photocopies of T.S. Eliot's *The Waste Land* into sustainable soil which he used to create the work *Eliot's Note Books*. These works take the soil in its various stages of transformation and reconstruct paper from it. In this manner each page of each notebook contains the words of *The Waste Land*. The notebooks evolve as a line of small sculptures implying an imagined line of editing. The final books of the work are entirely constructed in a soil that, if placed in the earth, would support growth and sustain life.

To make a project in response to Eliot's great poem is a privilege and a responsibility. During this year-long period I have sensed a structure in the text that feels to me like a map of places we all know but rarely bring their borders together. It seems, at one and the same time, to fold many of our histories into the everyday; moments observed, and histories remembered. The poem constantly seems out of reach and mysterious whilst giving just enough to keep our references to our own lives. It is a work that seems to possess a geology that layers the temporal, corporal and spiritual into its soil.

John Newling creating soil from shredded printouts of the T.S. Eliot's *The Waste Land*

Eliot's Note Books, 2018
Eliot's Soil, 2018
Installation at Coventry Cathedral

Eliot's Soil, 2018
Bags of soil made from paper, Butcher's Block, germination pots, paper bags
Installation at Turner Contemporary

Eliot's Note Books, 2018
Paper made from soil, rubber bands
Installation at Ikon Gallery, 2020

Eliot's Last Draft, 2018
Paper made from soil, Elastoplast, pressed and gilded flax seedlings

Art in co-relationship with nature

Anne Douglas
& Mark Hope

What defines an artist as an ecology artist?

How do we (humankind) re-imagine our place on the planet? How do we evolve ways of living that work with the grain of nature rather than seeing nature as something to be tamed or dominated? The enduring challenge is one of imagination. As we enter the 2020s, we are undergoing a profound rethink of what we value and how we organise ourselves, a process that spans the physical, metaphysical and spiritual.

It is clear that artists dealing with ecology are not 'ecology artists' in the way that 'portrait artists' deal with portraits or 'landscape painters' deal with landscape. Subject matter and materials are still important but social, environmental and political issues may be integral to the work and its generation.

> Ecological art is sometimes 'useful' and sometimes 'political', sometimes drawing attention to other ways of experiencing and understanding the world and our place in it, and sometimes involved in restoration of ecological systems. What might distinguish it are questions about who the aesthetic is for: Eco-artists imagine that the aesthetic could be for all the living things involved in the work, not just for humans.[1]

John describes his work as "in co-relationship with nature". How is this idea evident in the work? Instead of consuming materials to produce the work of art, John harnesses the energies of nature's processes (composting, seeding, planting, growing, flowering, dying, staining, bleeding). His process is generative, frequently leading to multiple elements within the same work. Importantly he imagines social and cultural domains to be part of the natural/environmental domain. Indeed a theme of his work that has spanned his entire career is exploring what it is to be human in relation to the *organic*, the *cultural* and the *etheric*. We might define the *organic* as related to living systems, the *cultural* as related to the way human beings organise themselves and make meaning as interdependent beings, and the *etheric* as the vital, life sustaining force present in all natural or living processes.

This is an intrinsically holistic practice in the sense that particular aspects of the subject matter are studied separately *not* to draw conclusions about the subject itself but in order to shed light on the whole. Given the interdependent nature of our world, the whole of something is rarely visible to us; indeed, defining what "whole" means is generally elusive as we live in a complex world where every element and system depends to some extent upon neighbouring systems. John's work usually starts with exploring a question or issue that opens up a line of inquiry into this complexity: What does a 21st century Eden look like? What do we do to make ourselves feel happy? Is guilt formed by nature or nurture? This is, at heart, an open-ended form of experimentation. One might even describe it as improvisation in the sense of creating new openings in our habits of thought. In this work, improvisation becomes a means to imaginatively join with those forces that sustain life, becoming part of life's continuity, imaginatively grasping what it might mean to keep going.

As in all forms of improvisation, freedom is subject to constraints. John's work deeply observes nature's systems, for example the manner in which a particular plant responds to the specific conditions of growth, a process that he meticulously documents as a personal relationship between himself as carer and the subject of his care. *The Lemon Tree & Me* (2011) is a work in which John cared for a particular lemon tree in this way over 688 days.[2] *The Lemon Tree & Me* could be read as a process of remediation that sought not only to extend the life of the tree, but to learn what was involved and transfer this understanding through sensitive documentation and analysis.

There is therefore more that is implied in the relationship of art to ecology than subject matter and materials; a different quality of relationship in which art emerges out of a dialogue or conversation, a form of co-production or as John describes it, a co-relationship with nature.

In exploring this further it may be helpful to define what we mean by 'ecology' and 'art'.

'Eco' comes from the Greek 'oikos' meaning 'home' or dwelling. Ecology shares this root with 'economics'. By tracing back to an etymological root associated with dwelling, 'economics' could mean the management of resources in relation to survival in the place that we call home. Ecology by association may be defined as the dynamic interrelationship between living beings within natural systems that continuously evolve and transform the environments in which they are situated (their 'home'). An ecology artist might draw out, uncover and reveal such dynamics, experiencing them by developing awareness of our entanglement with every component of our environment, a process in which we, the 'audience', become 'participants'. Ecology art is a dialogue of sorts between the artist, the natural world and ourselves.

The etymology of art is generally held to come from the Latin 'ars' meaning art, skill or craft but the Latin probably came from a much earlier root. The ecology artist, David Haley, reaches further back to some of the earliest human writing, the Rig Veda. 'Rta' is an Indo-Aryan noun/adjective meaning the dynamic process by which the whole cosmos continues to be created.[3] He has introduced this concept into policy thinking, not least CIWEM's 2009 policy position paper on art and ecology.[4] From this perspective, the whole of the cosmos including Life on Earth can be regarded as an artwork-in-progress. This might be true at every level and scale – at the level of individual life, family life, social life, work life or at the scale of the planet, the solar system or the Universe.

From these perspectives, the ecology artist works through art to explore different aspects of Life on Earth. In one sense, an ecology artist is similar to a scientist in using skill, observation, knowledge, logic and rationality to build a picture of how the world works. But the artist is able to draw on other skills that are denied (by self-imposed rules) to most scientists; artists can freely employ insight, intuition, imagination and emotion, enabling their ideas to reach beyond current beliefs and orthodoxies.

We have until relatively recently been accustomed to think of art and ecology as quite distinct and we frequently privilege scientific knowledge when addressing ecological problems. As our understanding of the natural world deepens, we are coming to appreciate the profound interrelatedness of all living systems, and hence the dangers inherent in drawing definitive conclusions from the study of isolated parts of nature. Ecology artists can help us imagine a different relationship between science and art.

How might these interrelated realms of art, science and ecology encounter each other productively? We explore this question through two particular bodies of work: the *Blackberry Picking* drawings *1, 2, and 3* (2013) and the *Waste Land* works (2017–present).

Blackberry Picking drawings *1, 2 and 3* (2013)

The *Blackberry Picking* drawings *1, 2 and 3* (2013) are celebrations of going to a particular place on the banks of the River Trent where John planted nine beech trees in 2009, known as *The Clearing*. The beech trees will eventually form a dome, creating a place where people can gather, recite poetry or simply be together. All around the space are blackberry bushes and each year Ann (John's wife and life long supporter) and John go to pick blackberries.

> The works came out of a mapping of that experience. They contain physical imprints of the blackberries; the dots you can see on the paper are the real plants. Some of the plants were allowed to sit on the paper for over a year so you have a sort of shadow of the plants. On top of that shadow, I started to do drawings about their geometry, both from *The Clearing* and from the blackberries themselves. Then there are a number of stains from the blackberries and from blood off my hands from picking. They work as documents of a moment; referring back to the picking.[5]

On top of each 'map' is a clock-like motif made from *Moringa Oleifera* trees that have been dried. Each map contains two stalks of Moringa that meet near the centre of the map like the hands of a clock, and each stalk has a nib at the far end of it. Across the ends, forming a cross, are gilded plasters. The layering of materials and process in relation to the passing of time carry a wealth of deeper meanings. The passage from Moringa trees to nib illuminate the idea that human evolution has gone from roots in the soil through to a very sophisticated notion of language. Paradoxically this trajectory of evolution has also brought about our progressive separation from nature. The nibs indicate the flowering of human consciousness into literature, poetry, philosophy, logic and science. Of course this flowering of consciousness has a dark side and it may not be accidental that of the six nibs in the drawings, three point vertically downward, one vertically upward and two laterally. Close to some of the nibs, gilded sticking plasters point towards our tendency to address symptoms rather than causes and, perhaps an even worse habit, to gild our 'solutions' to pretend that they might be permanent solutions rather than temporary patches. The gilded plasters also embrace deeper references: they form a cross with the Moringa trees/nibs; the cross is a very ancient symbol that represents the human condition as standing between the immanent (represented by the horizontal) and the transcendent (the vertical). The gilding itself has many biblical references to notions of value and very often, false value, as depicted in one of John's earlier works, *Currency and Belief* (2003)[6] that features a sculpture of the Golden Calf with the word "Mine" engraved into its side.

Each *Blackberry Picking* drawing is differently inflected. *Blackberry Picking 1* holds the trace of *9 Twists of Nature*, a theme that John has been working on since 1978. It is a drawing that follows a structure of a twisted ladder but using radii with successive rotations of 40 degrees relative to an axis, an experimental reflection of nature that bears a remarkable similarity to DNA. *Blackberry Picking 2* picks up on John's love of spinning through carefully drawn gyres – spinning cones, which can also be derived from *The Nine Twists of Nature*. This drawing is about the environment of the blackberry plants themselves, mapping the sunlight as it happened on the second day of picking through traces of objects across the sky and the spinning of a gyre.

> The two nibs meet at an exact point of a circle drawn when the sun was 6 degrees below and 6 degrees above the horizon – the liminal time at dawn and dusk when light becomes hazy and magical.[7]

Blackberry Picking 3 looks at the geography of the place itself in the clearing where the blackberry bushes are growing.

> I was plotting out blocks of land as I saw it, or rather remember it; so it's a more literal map still using the 40 degree twist; hence you get a circle consisting of nine elements. The nine elements are to do with organic, cultural and etheric; within each of those main areas that we exist in, there are sub-divisions of o-c, o-e, o-o, c-o, c-e, c-c, e-o, e-c and e-e. Kind of like a map which is in my head.[8]

John tells us that the drawings operate at multiple levels as works of art, as historical records – a documentary of process, as poetry, as geometry, as metaphor and as a search for meaning. They are part of a work-in-progress that originated with his early drawings of *9 Twists of Nature*.

What might we make of these drawings as a contribution to the field of ecology art?

Blackberry picking is a practice of foraging, of being outdoors and of harvesting the abundance of nature. It is about time and continuity, a pilgrimage of sorts, returning to the same place at the same time of year, taking part in the cycle of growth, breathing the air, appreciating the ever-changing light, treading the soil and sensing the moisture that feeds the plant and expands its fruit. The *Blackberry Picking* drawings enable us to recognise our place as one small element of a complex living system. In creating the drawings, John is never illustrative but builds experience in unexpected layers, using the actual plants through their imprints taken over a year, the crucial measure of time in this context. This cyclical time of the seasons can be contrasted with the different clock-times within all three drawings, the allusions to biblical history and mythological time represented by the gilded plasters and nibs.

The layering holds different ways of knowing from the measured and geometrical, time layered into the structure of the plant, plotted out through its cycle of growth, through to the visceral flow of blood and juice that impregnate and stain the drawing's surface. Nothing in these drawings is over determined. We wander through the layers of this open-ended exploratory form and come to see the blackberry picking in a wider context, to create a movement from the visceral and physical through the conceptual to the metaphysical, seeing the extraordinary in the ordinary.

Our engagement with the work is therefore an iterative and dynamic process. As we uncover different layers within the work, we encounter different ideas and memories in our mind. To engage with John's work is to engage with our relationship with the natural world, and to recognise that divisions of history, poetry, storytelling, geometry, art and the rest are simply different keys into a world of inter-relationship and, ultimately, wholeness.

The *Waste Land* works (2017–2018) including *Eliot's Notebooks*, *Eliot's Soil*, and *Eliot's Last Draft*

John's *Waste Land* works appear very different to the *Blackberry Picking* drawings and yet they share the same concerns and operate at similar multiple levels. The *Waste Land* works are based on the eponymous poem by T.S. Eliot (1922)[9]. *Eliot's Notebooks* and *Eliot's Soil* formed John's response to an invitation to participate in an exhibition, *Journeys with the Waste Land*, Turner Contemporary, Margate 2018. *Eliot's Last Draft* was a later commission for the University of Warwick Art Collection also in 2018. It is worth noting the pioneering curatorial approach at Turner Contemporary, Margate. The exhibition was a culmination of three years of a sustained and long-term dialogue with local residents who formed the Waste Land Research Group. That group developed the entire exhibition by discussing personal connections between art, poetry and life. This approach is resonant of the quality of dialogue that also underpins ecology art.

Eliot's *The Waste Land* is considered by many to be the most important and challenging poem in the English-speaking canon. It was written in response to World War I and the fragmentation of society brought about by the war with its unprecedented scale and horror. Written in five parts, it addresses an overwhelming preoccupation with death and dying. Death is not just the loss of life, but also the meaninglessness and despair that affected the whole social fabric. Eliot himself suffered illness and despair during this period and the poem juxtaposes horrific and desolate descriptions with references to older stories and classical literature as if seeking solace; can these stories and great works shed light on the incomprehensible carnage of war?

In his notes about *The Waste Land*, Eliot acknowledges the influence of Jessie L. Weston's book on the grail legend: *From Ritual to Romance* (1920) as well as that of James George Frazer's *The Golden Bough*, a study of comparative religion (1890)[10]. The Grail tells the story of the Fisher King whose mysterious illness causes sadness in his whole kingdom. This situation that can only be redeemed by a stranger arriving at the castle and asking the question: "Whom does the Grail serve?". Parsifal (literally 'innocent fool') is the young man who eventually asks this question and cures the King of his illness but not before he has spent twenty years in the wasteland engaged in gruelling quests and battles.

Frazer's *The Golden Bough* opened the West up to Eastern forms of religious belief and ritual and *The Waste Land* contains multiple references to Buddhism and Hinduism as well as Christianity. In seeking meaning in the face of the death and destruction of war, Eliot recognised that other faiths had something to add to his Christian belief. The poem ends with the Sanskrit: 'Shanti, Shanti, Shanti' (Peace, Peace, Peace). As Eliot himself notes: "... a formal ending to an Upanishad. 'The peace that passes understanding' is our equivalent to this word".[11] It points to the mystery of existence and allows optimism in spite of the apparent destruction and desolation.

The Waste Land seems particularly relevant today given the continuing dramatic loss of biodiversity and the climate emergency. The poem describes a world in which all certainties have been shattered; both the society and the beliefs upon which it was based appear broken. It is a world in which the previous order has collapsed, necessitating a renewed search for meaning.

John creates a counterpoint to Eliot's *The Waste Land*, acknowledging the fragmentation and disorder in our current times while exploring nature's capacity for renewal. This is evident not least in the later work, *Eliot's Last Draft* which plays with time and the existential threat to Life on Earth resulting from mankind's development. The horror of environmental change is however kept at a distance. John responds by undertaking a close, painstaking observation of nature, creating the conditions for new organic growth. John's *Waste Land* works involve the almost heretical act of creating soil from hundreds of shredded photocopies of Eliot's poem. As with Eliot's poem, the process was started in April ('April is the cruellest month' is the start of Eliot's poem) and the soil was completed in December 2017. Thus John connects into the natural cycle.

The soil John and nature create results from a daily ritual: a slow process of carefully adding the shredded text to vegetative matter that was collected each day from leftovers from the family kitchen, coffee grounds in the morning and vegetable scraps in the evening, along with urine as an accelerant of the process of decomposition. The two compost tumblers that held the mix were checked each day, turned, facilitating a spinning action through which the air is allowed to circulate, helping to generate the necessary heat to decompose the text. Then more was added.

> I do see the soil as a melding of a great work of human language within the great work of natural process; both sustain us and both help us grow.[12]

In this way Newling consciously closes the gap between human and natural systems. He follows the natural cycle of growth, starting in the spring – April is only 'the cruellest month' if you feel disconnected from nature and experience nature's spring revival in contrast to your own sadness or depression. Language is absorbed back into nature and made material again, joining the forces that sustain growth, literally. The whole dynamic is brought into consciousness through the work as art.

In the installation in Turner Contemporary (March 2018) there were seven sacks of soil. Soil from the sacks was offered to people visiting the installation, encouraging the public to grow seeds in the soil at home. These unlikely, deceptively simple acts in the context of art, reverse a number of tendencies in contemporary society: the tendency to consume rather than engage responsively with cycles of growth, the tendency to acquire what we need for survival through purchase rather than through gift exchanges in which care for other living beings and for the environment, is at the core. They also help us to understand where things come from and how they are made, and enable us to make things themselves.

Using the soil, Newling created the second work entitled *Eliot's Note Books*. These works take the soil in its various stages of transformation and reconstruct paper from it. Each page of each notebook therefore contains words from the poem in the literal sense of being embedded in the material of the shredded copies. The notebooks evolve as a line of small sculptures implying an imagined line of editing. There are 434 paper sheets in the work reflecting the 434 lines in the poem. The final books of the work are entirely constructed in a soil that, if placed in the earth, would enable growth and sustain life. Accompanying this work in the gallery was a film of the evolution and processes involved in the production of the soil and the processes of returning the soil to paper. Newling says:

> To make a project in response to Eliot's great poem is a privilege and a responsibility. During this year-long period I have sensed a structure in the text that feels to me like a map of places we all know but rarely bring their borders together. It seems, at one and the same time, to fold many of our histories into the everyday; moments observed and histories remembered. The poem constantly seems out of reach and mysterious whilst giving just enough to keep our references to our own lives. It is work that seems to possess a geology that layers the temporal, corporeal and spiritual into its soil. A soil that speaks of a great poet / anthropologist of all times and none.[13]

In engaging with these works it is clear that for artists working with ecology, the process involved in creating the work is as important as (and integral to) the 'finished' work.

Let's return to the opening question of what defines ecology art in relation to John's work.

John invites us all to create art in co-relationship with nature. We can experience this in his love of repetitive activities of planting, observing, watering and harvesting; of deep listening and asking questions; of collecting, sorting and analysing data; of shredding text and composting, marvelling at nature's ability to transform composted matter into new life. These activities frequently have a sense of ritual, repeated actions that reconnect him with the natural cycles and human processes that have sustained human beings over millennia, but which have largely disappeared as everyday experiences for us today due to our increasingly mechanised and urban way of life.

John's labour produces works of art. As objects they carry a degree of autonomy, enabling us to make meaning and give form to ideas in society. He goes further. He opens up our imagination to new possibilities of what art itself might be/become. If we accept the definition of art as the cosmos regenerating itself continuously, then the ecology artist can be regarded as one who illuminates aspects of that process for us, drawing us into close observation and reflection which enables us to look at the world differently. In this way "Art can change the parameters of our thinking".[14] Applied to the context of our current environmental crisis, 'changing the parameters of our thinking' is perhaps essential in order re-imagine what it means to be human and imagine how we can work in harmony with nature to create a world which can sustain a rich biodiversity.

John Newling's work as an ecology artist gifts to us works which can engage us at many levels, opening our minds to mystery and wonder, and encouraging imagination and improvisation. In challenging our habits of thought, in particular our tendency to look for certainty, it may also give us confidence to embrace uncertainty and treasure mystery, particularly the many mysteries that remain in nature.

Anne Douglas is an Emeritus Professor at Robert Gordon University, Aberdeen.
Mark Hope is a trustee and co-founder of Woodend Arts Limited.

1 Fremantle, C (2015). *The hope of something different* [online]. Available from https://arestles-sart.files.wordpress.com/2016/01/chris-fremantle-the-hope-of-something-different.pdf

2 Newling, J. (2015) *The Lemon Tree and Me*. Nottingham: J.B.N Publishing

3 Haley, D. (2016) *A question of values: Art, Ecology and the Natural Order of Things* in ed. J. Brady *Elemental an arts ecology reader*. Manchester: Gaia Press Project, Cornerhouse pp 41-62.

4 https://www.ciwem.org/assets/pdf/Policy/Policy%20Position%20Statement/Arts-and-the-Environment.pdf

5 *Blackberry Picking* drawings: John Newling in conversation with Mark Hope, the Barn Collection, Banchory, Scotland, 2013.

6 Newling, J. (2002) *Currency and Belief*. Published to coincide with the exhibition *Currency and Belief*, Yorkshire Sculpture Park, 30 November 2001-6 April 2003.

7 *Blackberry Picking* drawings: John Newling in conversation with Mark Hope, the Barn Collection, Banchory, Scotland, 2013

8 ibid

9 Eliot T.S. (1940) *The Waste Land*. London: Faber & Faber

10 Frazer, J. G. (1890) *The Golden Bough: a Study of Comparative Religion*.

11 Eliot T.S. (1940) *The Waste Land*. London: Faber & Faber

12 http://www.john-newling.com/blog/2018/3/eliots-soil-and-eliots-note-books-turner-contemporary

13 ibid

14 In conversation with Richard Davey and James Peto, Nottingham Contemporary, 2013 [online]. Available from https://www.youtube.com/watch?time_continue=29&v=Pbiu7U_UN6s&feature=emb_title

Trying To Find Where Here Is

2019

Trying To Find Where Here Is was installed in Tocil Wood, adjacent to the grounds of Warwick University. It is a permanent work commissioned for the Warwick art collection.

Tocil Wood is at least 400 years old and archaeological evidence shows earthworks in the woods dating back to the first century AD. The work is installed without any compromise to the mature oak tree and effectively floats over the surface of the bark by being tied on with horticultural rope. *Trying to Find Where Here Is* references the artist's book of the same year, *Dear Nature**. The title of the work and inscription on the stainless steel sign refers to humankind's loss of a sense of place within nature.

The sign reads:

> *'Our shadows alone touched you trying to find where here is.'*

* The central tenet of Newling's book *Dear Nature* (p 130) is the concept that the current ecological crisis is rooted in the human species experiencing a loss of a sense of place and a connection with their immediate environment.

Trying to Find Where Here Is, 2019
Stainless steel

Our shadows alone touched you trying to find where

Frank's Field

2018

In 2016 I was invited to be resident as part of the Diep-Haven residency. I was asked if I may be interested in working with a co-operative of flax farmers based in and around Dieppe. My work for Diep started with me germinating wildflower and flax seeds, given to me by Franck Sagaert, a farmer from the co-operative. I was particularly interested in how the seed first germinated and how roots evolve under the soil. I was further interested in how Franck manages the co-relationship between himself and the natural ecology. I have worked on many projects that review my own co-relationship to a natural ecology, starting through the agency of growing a specific plant or plants. For this reason, I felt some empathy with Franck.

John Newling with flax farmer Franck Sagaert on the farm in Dieppe.

Bundles of harvested flax ready for processing into linen

Frank's Field, 2019
Pressed flax seedlings on paper

Frank's Field (floor piece)
2018

Frank's Field was one of several works that emerged from the Diep-Haven residency looking at flax farming, beginning with the growing of flax using seeds given to the artist by farmer Franck Sagaert. The work consists of 243 casts made of soil used for growing the flax, and text from the *Dear Nature* letters, a series of letters Newling wrote to Nature (p 130). In each tile are embedded elements of the flax plants themselves.

Frank's Field is a kind of conceptual carpet that has embedded in it the visual situations, texts and activities developed from growing the seeds Frank gave me.

Frank's Field (floor piece), 2018
Soil slabs, pressed flax seedlings, string, chalk

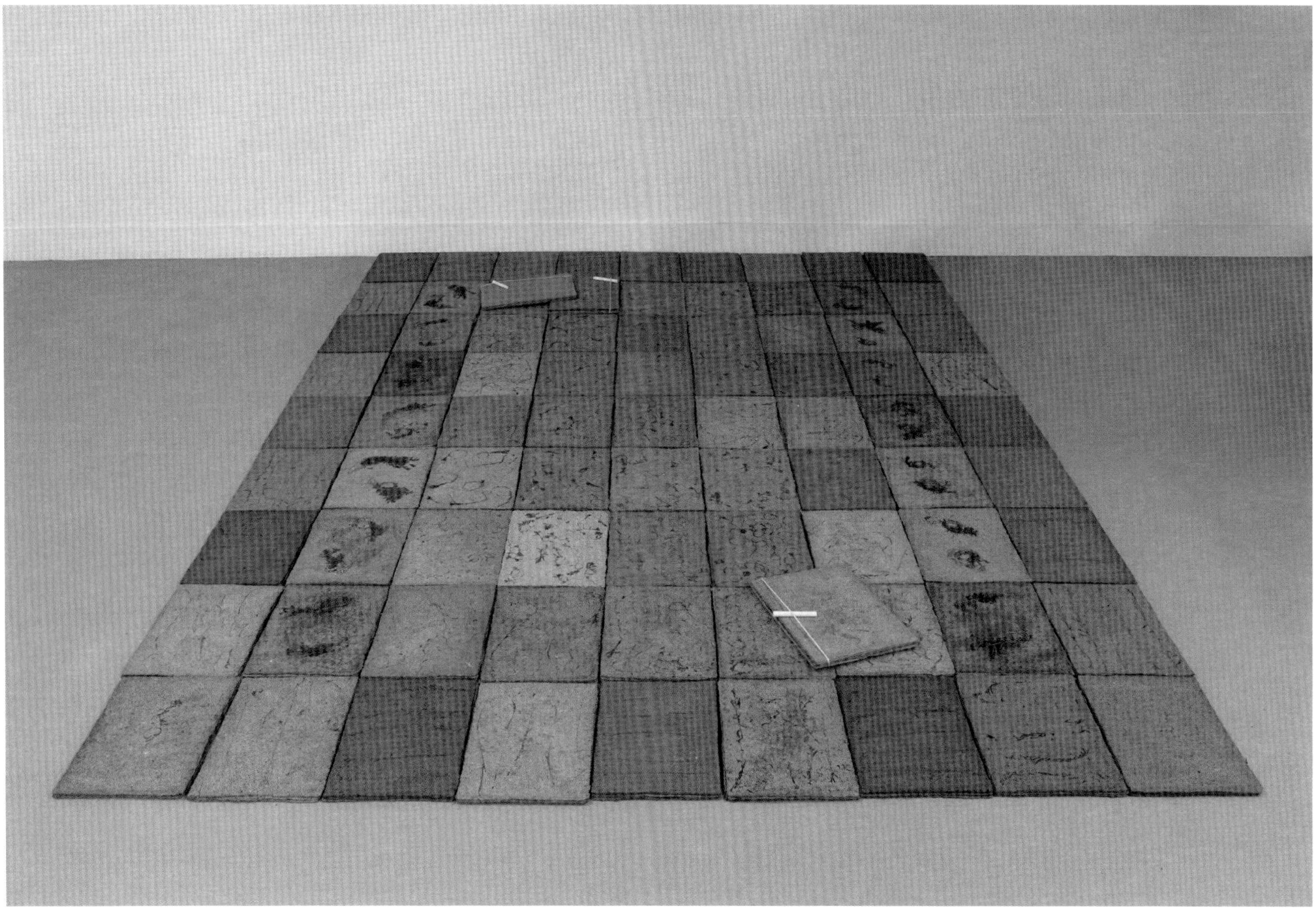

Dear Nature
2018

Following on from the *Dear Nature Letter* (p 104), in 2018 Newling spent 81 consecutive days writing letters to Nature. Without knowing the final outcome, over a period of just under three months, the artist rose at 6am everyday and wrote a letter to nature.

The series of letters echo the sentiment of letters written to a loved one but with a voice of humanity attempting to come to terms with its relationship to nature. The letters take you on a journey through the history of agriculture, economics, religion and capitalism to the present day. Newling describes the values that have underpinned human endeavour and shaped our world.

Printed alongside the letters are a series of images of pressed flax seedlings grown from seeds given to the artist by farmer Franck Sagaert, as part of the Diep-Haven residency.

Throughout his career Newling has maintained a writing practice that runs in parallel with his art making. Writing is often used as a tool by the artist to contemplate ideas contained within his work and frequently operating as the link from one artwork to another. *Dear Nature* is an example of how the artist uses writing as a performative act and a process by which to formulate new artworks.

Dear Nature describes that which all humans with a conscience must, deep down, feel right now about nature. The book offers a new state of mind for the future, rooted in balance and respect.

Pressed flax seedling

Pages from *Dear Nature*, 2018

John Newling
Midlands, UK
Nature
The Earth
21st March 2018
Dear Nature,
We have changed the climate.
We are under the weather.
We have changed our climate.
We are under our weather.
Yours,
John

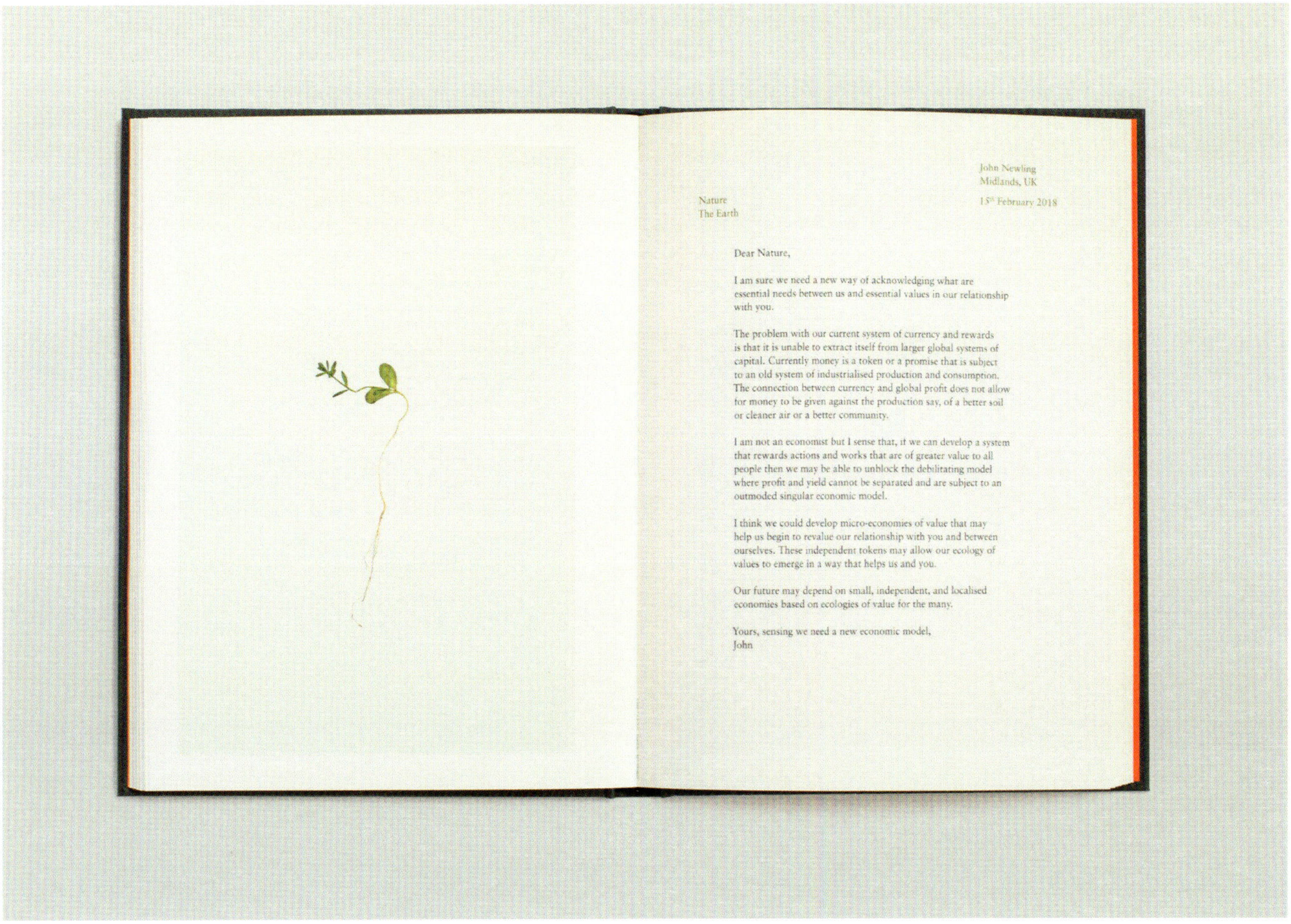
John Newling
Midlands, UK
Nature
The Earth
15th February 2018
Dear Nature,
I am sure we need a new way of acknowledging what are essential needs between us and essential values in our relationship with you.
The problem with our current system of currency and rewards is that it is unable to extract itself from larger global systems of capital. Currently money is a token or a promise that is subject to an old system of industrialised production and consumption. The connection between currency and global profit does not allow for money to be given against the production say, of a better soil or cleaner air or a better community.
I am not an economist but I sense that, if we can develop a system that rewards actions and works that are of greater value to all people then we may be able to unblock the debilitating model where profit and yield cannot be separated and are subject to an outmoded singular economic model.
I think we could develop micro-economies of value that may help us begin to revalue our relationship with you and between ourselves. These independent tokens may allow our ecology of values to emerge in a way that helps us and you.
Our future may depend on small, independent, and localised economies based on ecologies of value for the many.
Yours, sensing we need a new economic model,
John

Conversation

Richard Davey

Artist's studios come in all shapes and sizes. From cavernous art factories where the act of creation occurs on an industrial scale, to sheds, where the solitary artist finds creative escape at the bottom of the garden. Some are pristine and ordered, others are full of detritus, splattered with shards of paint, filled with half worked canvases, plastered in quotes, postcards and notes of ideas. They can be in warehouses or homes, communal or solitary, filled with noise or echoing with silence.

John Newling's studios, on the top floor of his Nottingham home, are places of quiet industry that were once small spare bedrooms. Through their windows he looks out at the world, staring through a microcosm of individual leaves, roots, tiles, bricks and wood into the unfolding macrocosmic depths of the universe. Although the two rooms are of similar size, they play very different roles in his practice. In the studio at the front of the house he sits at his computer, watching, writing, reflecting, thinking, remembering. He gazes out at the detached houses opposite, their front lawns bordered by mature hedges. Parked cars line the wide street, while mature trees punctuate the grey pavements with an organic disruption that dissolves edges and neat order. Looking down at the tree across from his front gate, Newling's gaze probes its undulating root zone, where asphalt and bark intersect. He searches for the presence of life, remembering the pungent perfume of the herbs he once planted and feeling the ghostly presence of the soil that impregnated his fingertips as he undertook the acts of guerrilla gardening that have transformed this suburban borderland into an intersectional space. It was here, sitting at his computer, contemplating this edgeland between wild nature and humanity's 'concrete' order, that Newling wrote his epistles to the natural world, *Dear Nature*.

The words of *Dear Nature*, framed and shaped in Newling's mind, reaching out to the silent other, yearn for a response, which never seems to come. They lie on the page listening for forgiveness, searching for understanding. But they speak in words, lying static on a page, moving only when spoken in the mind or out loud, whereas nature speaks a language that humans have long since lost the ability to recognise or understand.

Across the landing is another studio. Files and folders fill filing cabinets and shelves, holding the memories of past projects and art works. Like a monastic cell, this is a place of silence and purpose. Everything is neatly arranged and has a place, with nothing extraneous or cluttered to distract his attention. Taking up much of the room in this confined space are two tables covered with the implements and products of creative activity, for it is here that Newling moves from the interior world of concept to the act of making, turning ordinary objects into extraordinary things. Sometimes what emerges are sculptural forms but increasingly he has turned natural objects into paintings, although there is little evidence in this ordered space of the mess and debris that you usually find in a painter's studio.

Looking out of the window of this north facing room, the contained, intimate space of the crescent gives way to a vista of terraced houses, which dip away and disappear from Newling's gaze, before rising up and ending in the distant prospect of Nottingham Prison, whose brick walls delineate the horizon. Immediately below the window Newling can look out onto his back garden, where carefully tended lawns, trees, beds, plants, and a greenhouse provide a green lung to mitigate against the urban claustrophobia, where the ordered ranks of terraced housing dissolve into an anonymous haze of bricks, tiles, slate and concrete.

From his upstairs vantage point Newling stands and observes, a voyeur waiting for Nature to reveal its intimate secrets and expose its mysteries. He watches as trees bud and come into leaf, each year growing a little bigger, and then he marvels as they lose their foliage in a final moment of ecstatic display. He is a witness to growth and decay, life and death. He stands mesmerised as branches dance and etch the air with their wind-borne choreography. Yet despite the attentive intimacy of his fascinated gaze, he remains blind to the chemical conversations and innate electrical and pheromonal responses that allow trees to communicate, bond and warn each other of danger. He can never witness the co-operation of the forest as the individual supports the whole, nurturing the sick and dying or sharing news of predators so that the rest of the community has time to prepare and protect itself with chemical responses. He may listen attentively for this complex root language but in the end, he can only imagine the sound of woody tendrils pushing down through the soil, probing deep into the earth for reservoirs of water or reaching up to the sun, drawing concentric rings to delineate their search for light.

Trees and plants have been a source of inspiration throughout Newling's career. He has cultivated a crop of walking stick cabbages in his back garden and planted a clearing of trees by the River Trent. He has wrapped branches in printed prayers and petitions and embedded messages in tree trunks. He has watched by video link as a Pinot Noir vineyard transformed the interior of a redundant church in Kent and later, he learnt how to grow moringa trees in hydroponic tents. He has pressed leaves, painted them and gilded them. He has stuck them together and punched Communion wafer sized holes through them. He has used their forms to construct an alphabet, writing love letters to them in this leave-moulded script. He has turned paper into soil for lemon trees to grow in and watched transfixed as the tree bore fruit, and in the cavernous cellar studio beneath his house he continues to grow and nurture moringa trees.

But despite Newling's fascination with and abiding respect for plants and trees, these works have inevitably been only a one-sided conversation, emerging from an idea already formed in his head. Nature has inspired him, but it has also been his silent partner. He has shown its wonder, revealed its miracles and explored its potential, but he has done so through a language shaped by mind and aesthetic norms. Yet Newling's constant contact with these living things is slowly opening up a new conversation as he becomes exposed to their distinctive voice.

When Newling tasted the wine pressed from his Chatham grapes and swirled it around his tongue, he was no longer watching the grapes grow, ripen and fill the empty nave remotely through cameras, he was absorbing their distinctive terroir into his body, becoming, without realising it, a container for light, time and place. He has squeezed 'soil' between his fingers, feeling it permeate his skin, inking his palms with nutrient-rich life, his hands became rooted in the earth. He has picked tiny moringa plants and giant cabbage leaves to press and glue together, holding them in his hand and marvelling at their structure and form, transfixed by the natural calligraphy that delineates time and growth. As he has watched, picked, tasted, felt, probed and smelt, the language of trees has seeped through his skin, and been absorbed into his DNA, forming a persistent wordless whisper in his body that has gradually taught him to be attentive to nature's distinctive voice.

So now, when Newling goes downstairs and walks into his garden, he has stopped looking with his eyes alone and thinking with his mind, he has started to see through touch. He has learnt to smell the subtle aroma of the leaves and taste their surface with his eyes. And as he holds them between his fingertips, their skin touching his skin, the whorls of their fingerprint ridges interlocked with his, feeling their textures and investigating their crevices, he is drawn into a different time zone, where life is measured in centuries and millennia and time is measured in rings rather than straight lines.

He bends down and picks up leaves that have fallen to the ground, their job done, their life returning to the soil. Collecting twenty different leaves he takes them back upstairs and carefully presses them, squeezing them between the leaves of a press until they are transformed into a flat shape. He then places sixteen of them in a grid on a sheet of folded Arches watercolour paper, within which he has placed letters, private missives, which like the coins he once buried within the roots of the trees that formed his *Clearing*, will remain hidden, a secret bond uniting creator and created.

He then gilds the whole paper with either silver, gold or copper leaf before removing the leaves, leaving behind their ghostly trace. He then floods the gilded surface with paint, probing it with brush and fingers, gliding across slippery metal, stumbling across paper left exposed by the removal of the leaves. As the bristles of Newling's brush flicker across the surface, pushing into and investigating the spaces left by the leaves, tasting these voids with their paint-laden presence, he discovers a choreography of growth. Lines emerge and dart into being, twisting and swirling, tracing through the movement of his arm the unfurling of trees into their seasonal habit, as they become laden with leaves. The ordered grid which Newling initially imposed disappears, solid forms dissolve leaving only a whispering, shimmering, effervescent trace lingering in their pungent atmosphere, an elusive hint at a language yet to be fully learnt.

In other 'leaf works', like *Nyman's Language* and his Soil Books, Newling leaves the leaves on the surface so that they stand out as definite shapes. Removing them in the *Library of Ecological Conversations* means the surfaces are less assertive and defined. These are not statements, but the beginning of a conversation, autonomous objects tracing a fugitive exchange. They form a primer for our conversation with nature, the hesitant first steps at learning a new way of communicating with the other, defined by listening rather than stating, touch rather than speech, intuition rather than imposed ideas. When Newling builds up the layers of colour on each painting he doesn't know when he will stop, only that he will stop when it feels right. He no longer knows what he has learnt, only that the exchange has changed him, transforming his vision so that when he looks at the world now he sees lines of plant growth rather than alphabets of human knowledge, colours as signs of light and growth, rather than symbols of meaning and memory. The making, thinking and growing, which once occupied separate studios are now merging into one, so that growth is making leading to thinking, leading to making leading to growth, in the seasonal circularity he has learnt in his conversation between the 'leaves and me.'

Richard Davey is an author and essayist of multiple books including 'Spinning – Nature, Culture and the Spiritual in the work of John Newling' (Nottingham Contemporary), 'Anselm Kiefer' (RA Publications) and 'Anthony Whishaw, Works on Paper' (Beam Editions).

Dear Nature (Exhibition)

Ikon, 2020

Dear Nature, the exhibition shown at Ikon in spring 2020, takes its name from the artist's book published in 2018 and brings together a series of works that were made specifically for the show and work from the previous three years.

The Library of Ecological Values, composed of 36 framed works that dominated the space, continues his exploration of nature and language, which began with the project *Nymans Language*. To make these works the artist worked board surfaces with paint, integrating leaves into the process, then removing the leaves to reveal ghost-like outlines of these natural forms. Following the lines and grids of a script, the works are a collaboration and conversation with nature. Newling further explores the relationship between language and nature in the *Soil Books,* a series of books made from sheets of 'soil paper' produced by the artist, and invites the viewer to read a script of leaves.

To complement these works, *365 Days and 50-Million-Year-Old Leaves* (2019) and *Burnt Sticks* (2019), are a series of sculptures composed of stacks made from sticks the artist collected on his daily walks near his home in Nottingham. The sticks are stripped back and painted black with white tips to create magic wands. The magic wands are both a metaphor for radical transformation in terms of human / nature collaboration and continue the artist's philosophical interest in the concept of 'faith', a theme that featured in earlier works in the 1990s.

Dear Nature
Installation view, Ikon Gallery, 2020

A Library of Ecological Conversations – Leaves and Me

2018–2020

For almost a decade I have been building a body of works that are two-dimensional. Each work attempts to review, and respond to, conversations between place, nature and me. Each work articulates a visual language that has evolved through my close observations and research into the growing of particular species of trees and plants. The work has material and conceptual processes that reflect each conversation as a situation seeking to construct a language that talks of place and ecology.

A library of Ecological Conversations – Leaves and Me are works I make to try to understand the multiplicity of languages that bridge between art and nature and us. I see them as letters of a sort between me and nature and nature and art. Each work starts with a selection of leaves, roots and flowers that I have in a kind of library of such elements that has evolved over the years and makes reference to many of my growing projects, with material gathered from many different places.

Each of the works consists of three imperial sheets of Archers hot-pressed paper which is folded and pressed to approximately the dimensions of foolscap; a size often associated with letter writing. Each of the three folded works are then glued and pressed to form the surface on which the language can emerge. The collected leaves are glued to the surface and then a layer of gold, silver or copper is applied to cover the whole surface. The leaves are then removed leaving ghosts of their shape. The work then becomes a conversation as paint and lines are applied and expressions of what the work is saying are developed until it, and me, are exhausted and the work is complete. It is a correspondence of a kind. These are intense cartographies of languages formed within the processes of the work itself.

A Library of Ecological Conversations – Leaves and Me
Installation view, Ikon Gallery, 2020

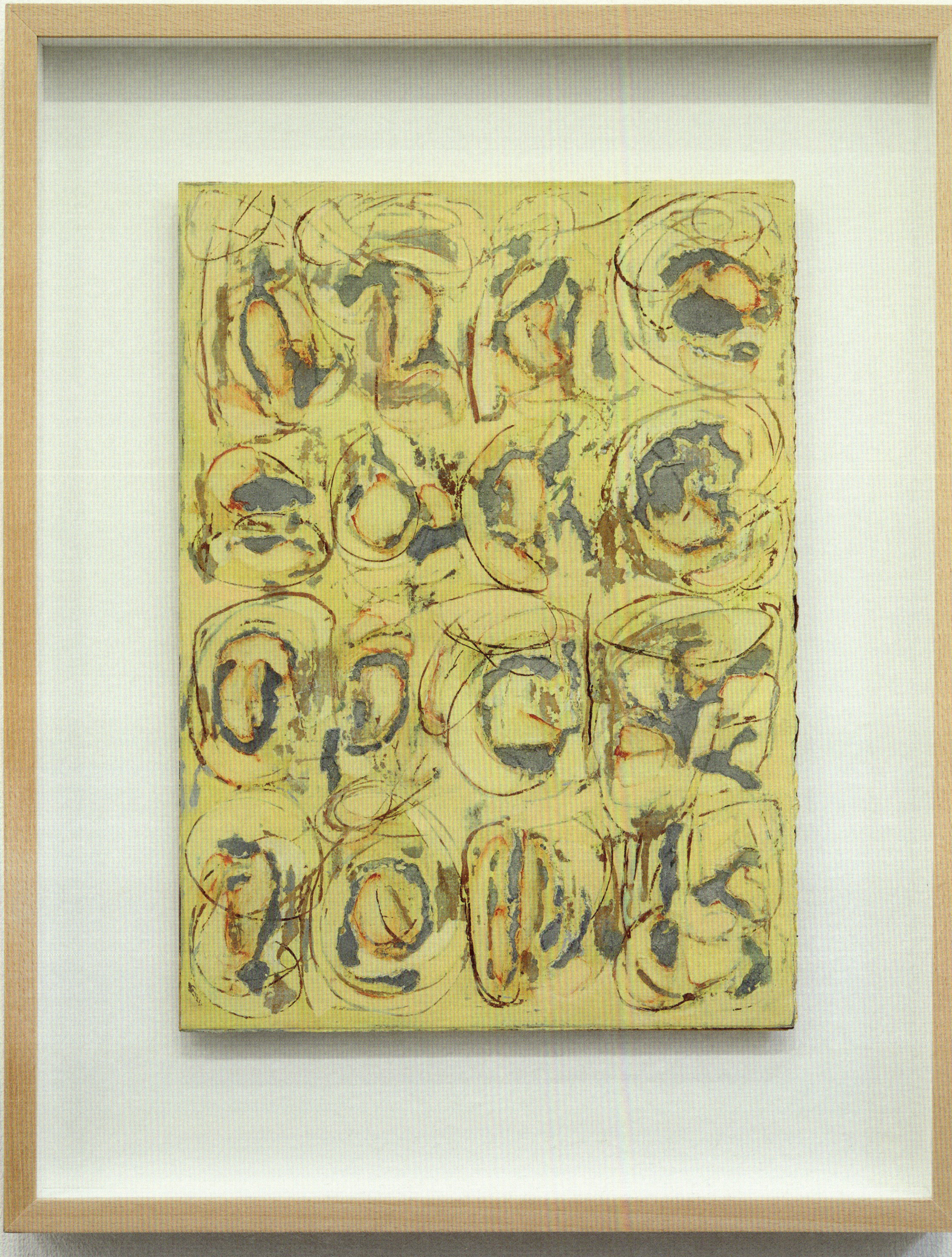

Soil Books

2020

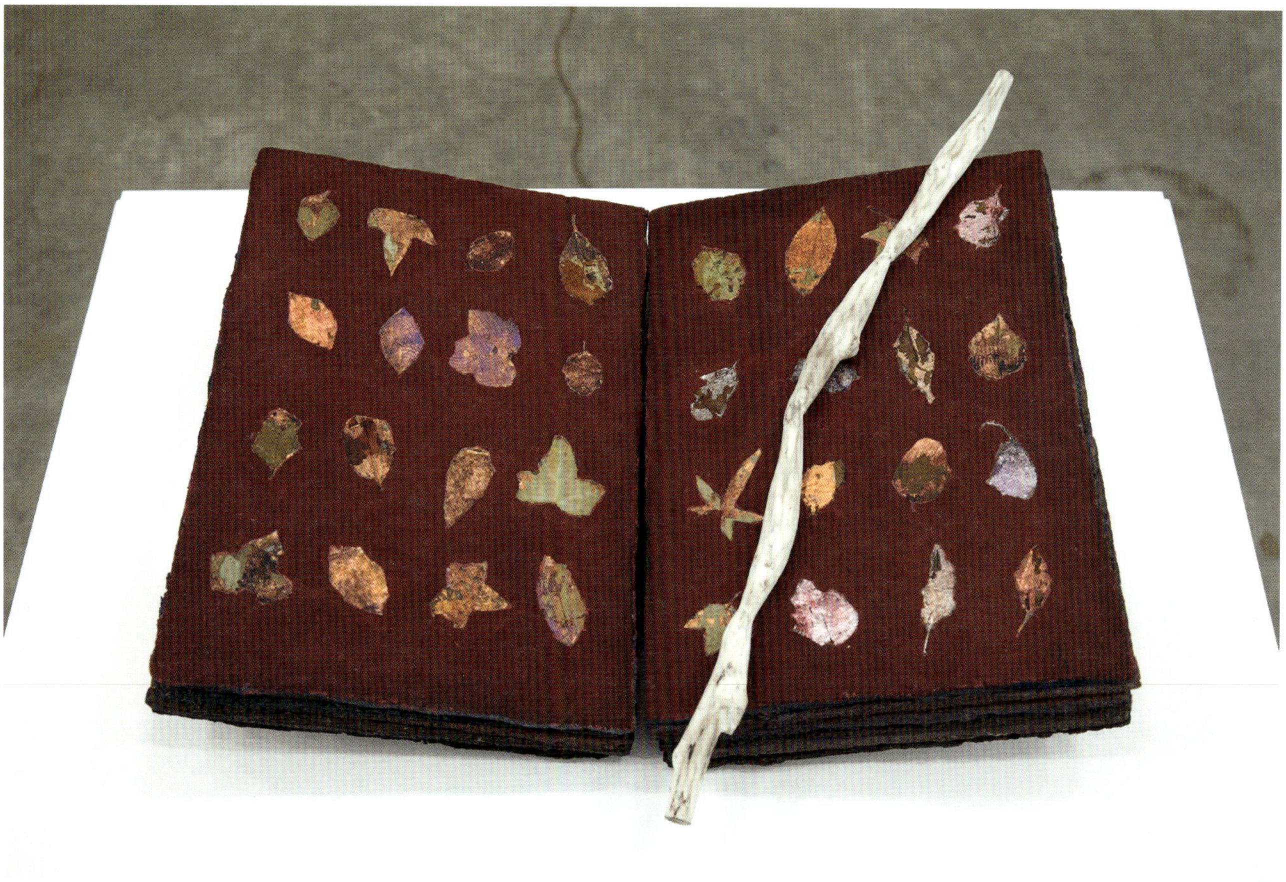

365 Days and 50-Million-Year-Old Leaves

2019–2020

I walk a lot. I walk for many reasons. Often on my walks I see the wonderful maps ascribed by the falling of sticks and leaves from trees. I find these natural configurations beautiful.

I collected sticks on each walk I took over a period of a year. Each stick I collected had the bark removed and was painted in the manner of a magic wand. Back in the studio each sculpture was slowly constructed as I stacked bundles of 7 sticks (days) together. Each work has approximately 365 sticks in it. The sticks are painted in watercolour so when it rains the magic washes out leaving the real magic, the sticks themselves, behind.

It was a lovely coincidence that when I started this work our grandson, Stanley, then aged 1, also loved picking up sticks and carrying them with him. If it was good enough for Stanley it was good enough for me.

365 Days and 50-Million-Year-Old Leaves, 2019
Sticks, watercolour, inkwell, feathers, fossils, soil

Burnt Sticks

2019–2020

Burnt Sticks, 2020
Sticks, watercolour, rubber bands, timber, soil

Dear Nature (Film)

2019

In 2019 John Newling commissioned artist and video-maker Reece Straw to interpret the *Dear Nature* book into film. Reece recorded a range of different people from the local community reading the letters which were then set to video footage of nature captured from the local environment. The result is a meditative feature length film that visualises the philosophical messages of the original book. Public participation is a recurring feature of many of Newling's projects. The *Dear Nature* film is another example of how direct participation from people within the local community is used by Newling as a method for engagement. The concept of sharing and exchange between the artist and the public is conceptually embedded within the work.

Dear Nature (Film), 2019
4K Video
In collaboration with Reece Straw

So Sorry
2020

So Sorry, 2019
Stainless steel

We Have Signed Our Names in Your Soil, So Sorry, 2019
Rubber stamp, mirror

John Newling
Midlands, UK

27th March 2018

Nature
The Earth

Dear Nature,

A lament.

We were lovers; your unknowns folded us in fear and wonder.
We signed our names in your soils lost in our clearing and clearings.
Our shadows alone touched you trying to find where here is.
Wanting here to be elsewhere we became estranged.

I am sorry but, perhaps we are moving towards knowing what here means.

Yours,
John

From My Garden

2018

From My Garden, 2018
Acrylic, copper gilding on paper

Light and Day
2018

Light and Day, 2018
Acrylic, gilding on paper

Small Sculptures and Monuments

Jonathan Casciani

Newling has often worked on a monumental or beyond a monumental scale. *Chatham Vines* (2006) the hydroponic vineyard the artist installed in a disused chapel in the Kent town of Chatham, transcends the concept of a sculptural installation or a site specific work. The church and the hydroponic vineyard were symbolically bound together. The church is not just a contextually relevant location, but an essential component of the artwork, and the vines activate the space to make the whole of the church an artwork.

The artist has made works that are monumental in scale but these works have never reflected the qualities of great monuments. Monuments are designed to wield power over those that stand beneath them, they are didactic and often a device for facilitating ulterior motifs.

For *Value; Note, Coin and Eclipse* (2011) the artist transformed the stalks and leaves of Jersey Kale cabbages into a grid of 27 framed works that stretches approximately eight metres across a gallery wall. The gridded work stands as a new monument for the twenty first century, one that symbolises the importance and enormity of natural systems. In contrast to monuments that usually stand as single entities, this work is composed of a series of individually framed works. These intimately crafted components can be read individually and collectively like the pages of a book, one gilded cabbage leaf at a time. To 'read' a large Newling work is not to consume a singular monumental message, it is to pursue knowledge on a more intimate scale.

Newling has made a number of small works since the *Ecologies of Value* exhibition (2012) and has continually produced small sculptures composed of different objects that are intimate in scale, but no less powerful for their modesty.

Books, balls and bowls made of soil, exposed roots, sticks, leaves cut into archetypal house forms, sticks made into magic wands tied in bunches with fossils and feathers. Like the plant leaves that form the alphabet in *Nymans Language* (2017), these objects are like characters of an alphabet that come together as a new 'language' that can be used to express ideals and values.

The bowl is a recurring device that is used throughout these sculptures. In Newling's case the bowl is delicately made from a soil that is constructed from paper containing a symbolic text that has been shredded and composted or soil dug from the artist's garden. A bowl is a carrier for food, a symbol of generosity among humans, a symbol of sharing. The delicate nature of a bowl made from soil expresses the vulnerability of the act of sharing and how generosity itself, in our highly industrialised environment is under threat.

Generosity is always a gesture of trust and the artist trusts the complex fragility of the soil, to retain its form and be a carrier for other objects. The soil bowls are an expression of the need to trust rather than control, as a value and an ideal.

> 'In the studio I find myself constantly cleaning the space I am working in. This must have something to do with control in a very uncertain world; a kind of transfer from obsessive washing of hands to the place I work in. The new works are housed in bowls made of soil. In the case of a couple of the bowls, pigment has been added to the soil; a sort of human/nature collaboration. With these works the soil has been dug from our garden. The soil is sieved and mixed with water and a binding agent. The soil is then cast into a mould. It takes a minimum of 3 weeks for the soil to properly dry and set.
>
> 'As with many of my works they are a mixture of natural materials collected from where I am that become subject to processes where the object of the work slowly evolves. They are a co-relationship between nature and me; an ecology of forms, materials and values as a way of better understanding and seeing the world. Each work teaches and directs me as material and conceptual forms and thoughts emerge.'

During the lockdown due to the Corona Virus pandemic, Newling collaborated with horticulturist and gardener Alys Fowler on a series of letters to one another, sharing their experiences of their daily lockdown walks that are published on Ikon Gallery's website. Morning walking is a daily ritual for Newling. As a student in the mid-seventies he made works from his experience of a daily walk along the canal towpath between Burslem and Stoke-on-Trent. These towpath walks were to inform many of his future thoughts and works.

It provides the artist with a time for contemplation and sets a foundation for the day ahead. During this time the artist created a series of artworks that continued his use of small objects constructed from raw natural materials:

> 'On each walk I collect a stick from the floor of woods and streets. Back in the studio I cut the stick to length, remove the bark and paint it a shade of yellow. For some reason this habit seems to help me; giving a kind of purpose to the walk. I can also recall where each stick has come from. In the studio they seem to annotate a rhythm to the days; almost sonic in some ways. To date I have some 47 sticks; I suspect many more to come before this lockdown is lifted.'

These sticks were bunched together and tied with string becoming an archive of nature. Like points of navigation on a map, the sticks are a way of recording the location and connecting with that locality. Newling has long been an advocate of 'trying to find where here is'. What he infers through this is that an awareness of our local environment, where we are, is important. Newling would argue that knowing 'where we are' means you are able to make the small-scale actions that will heal the environment. An early question of Newling's 'why do we need angels to need wings?' from the artwork *Questions, Ices And Places (Inserts)* (1999) is a marker of his desire to see our earthly values progress toward making a better place. A better place for Newling is a place where humans and the natural environment are constantly involved in a co-relationship where we learn more about the earth we inhabit.

After the bunched yellow sticks Newling went on to create three more sculptures whittling sticks into 'pencils'. A black pencil with white tips sits in a bowl with two soil balls, one gilded with gold leaf. Another bowl in a muted blue, carries a group of yellow sticks and a third sculpture carries 'pencils' in various colours. These objects are man-made but with nature dominating their method of construction and appearance. The objects symbolically present an equally balanced relationship within nature. The pencil is also an instrument for expressing new ideas or making new plans. The black pencils with the white tips reference the drawing instruments of the engineer, a tool for precise design. Coloured pencils are the spontaneous tool of the child or the artist, where ideas are formed without the need for precise application and a clear means to an end. Nature is simultaneously precise and spontaneous, and these sculptures show us that we need different tools and methodologies to co-exist within nature.

> '... This spring feels like no other I can remember. It is so overwhelming in its beauty; the whole environment seems saturated in blues of sky and yellows of the sun. Nature does not stop for us but we can stop for it...
>
> '... On each walk I feel the earth is getting cleaner and I think that, in some ways, all of this can be seen as a rehearsal for some of the conditions we will need to maintain if we want to survive as a species...
>
> '... I am starting to worry that all these wonderful experiences of the natural environment will be forgotten in our rush to regain where we were. It must be possible to see this terrible virus as giving us a lesson to better understand our co-relationship within the natural world...'

This co-relationship with the natural world that the artist talks of is expressed through the small sculpture that shows an archetypal house form constructed from layers of leaves, delicately balanced on a ball of soil, further balanced on a bottle of paracetamol moulded from soil. The small objects are playfully in dialogue. The arrangement has great fragility and it's long-term survival relies on trust and care by its custodian.

> 'These small-scale artworks are a new kind of monument, one that can be read on a human scale. They exist as expressive models of thought and materiality; hovering between our human ways and the natural environment. In part, they ask that we coexist in scale with a nature that we experience every day.'

Jonathan Casciani

Tillage Works
2020

Publications

JOHN NEWLING: *Nymans – A Language From the Garden*. Nottingham, UK: J.B.N. Publishing, 2016. ISBN 978-0-9542484-2-0

JOHN NEWLING: *The Lemon Tree & Me*. Nottingham, UK: J.B.N. Publishing, 2015. Limited edition of 175 copies. Published on the occasion of the conference 'Climate Change and the Common Good', St Paul's Cathedral, 23 April 2015, London. ISBN 978-0-9542484-1-3

JOHN NEWLING: *The Last Islands*. Nottingham, UK: J.B.N. Publishing, 2015. Published on the occasion of the opeing of *The Map Room of the Last Islands* at Woodend Barn, Banchory, Aberdeenshire. ISBN 978-0-9551397-8-9

RICHARD DAVEY: *Spinning – Nature, Culture and the Spiritual in the Work of John Newling*. (Peter Sawbridge, Ed.). Nottingham, UK: Nottingham Contemporary, 2013. ISBN 978-1-907421-06-8

JOHN NEWLING: *The Noah Laboratory; Constructing Soil*. Lincoln, UK: The Collection, Lincoln, 2009. The book includes a foreword by Lesley Farrell, the essay 'From my Garden' by John Newling and a critical overview by Dr Jonathan Willett. ISBN 978-0-9562377-0-5

JOHN NEWLING: *Make a Piano in Spain*. London: Wellcome Trust, 2008. Includes a forward by Ken Arnold and a contextual essay by Dr Jonathan Willett. ISBN 978-1-84129-079-9.

JOHN NEWLING: *Preston Market Mystery Project*. Preston, UK: 2008. Includes a foreword essay by Bob Dickenson, a DVD of the live reading event and an edited film of the project. ISBN 978-1-871575-27-9

JOHN NEWLING: *An Essential Disorientation*. SARP, Poland, 2007. Commissioned by the Polish Museum of Contemporary Sacred Art and published by SARP. English and Polish. ISBN 978-83-922964-0-9

JOHN NEWLING: *Chatham Vines*. London: ArtOffice, 2006. Includes essays by Newling, Isabel Vasseur, Andrew Spira, Peter Locke and an extensive interview between Paul Bonaventura and John Newling. ISBN 987-0-9543608-1-8.

JOHN NEWLING: *Writings by John Newling 1995–2005. Book One, Essays 1995–2004, Book Two, Transactions and Agreements*. London: SWPA, 2005. Prefaced by Johannes Birringer. (Two volumes). ISBN 0-9547 300-2-x

JOHN NEWLING: *Stamping Uncertainty*. London: SWPA, 2004. Includes essays by John Newling and Andrew Spira (Christie's Education). ISBN 0-9547300-0-3

JOHN NEWLING: *The Westonbirt Wishes*. London: SWPA, 2004. Essays by Newling and foreword by Samantha Wilkinson. ISBN 0-9547300-11

JOHN NEWLING: *Currency and Belief*. Bretton: Yorkshire Sculpture Park, 2003. Published by Yorkshire Sculpture Park in conjunction with the exhibition of the same name. Includes essays by Antonia Payne, Simon Herbert, Clare Lilley and John Newling. ISBN 1-871-480-41-8

JOHN NEWLING: *A Short Walk*. Public Art Journal, Volume 1 Number 6. p. 52–61, 2002. ISSN 1465-7376. Further published, 2003. *A Short Walk*. The Now Festival Catalogue.

JOHN NEWLING: *Prayer Not of Proof but of Love*. Worcester: Worcester Hospital Chaplaincy, 2002. ISBN 0-9542484-0-6

JOHN NEWLING: *1.0*, Video of the Performance and Installation for Making History at Stafford Prison. Stafford: Making History, 2001

JOHN NEWLING: *PLACE; Selected Writings*. Turnpike Gallery, 1998. Published in association with Newling' project *Weight*. ISBN 0-9529470-1-3

Exhibitions (2001–2020)

Please note all exhibitions are one person shows unless specified otherwise.

Tillage. 2020. Exhibition/installation at Ditchling Museum.

Dear Nature. 2020. Exhibition at Ikon Gallery, Birmingham. Curated by Jonathan Watkins and Thomas Ellmer.

Received Wisdom. 2020. Sunderland Museum and winter Garden. Arts Council Collection, national partner exhibition.

The Making of Landscapes. 2019. Exhibition at La Galerie Duchamp, France. Curated by Julie Faitot, Alice Mallet.

Trying to find where here is. 2018–2019.Warwick Art Collection commission for 3 works; the book 'Dear Nature', a two dimensional work and a sculpture work for a forest. Curated by Sarah Shalgosky.

Terra Firma. 2018. ONCA Brighton. Curator, Rosie Herman.

Criminal Ornamentation. 2018–2019. Curated by Yinka Shonibare MBE from the Arts Council Collection. (Group show).
Attenborough Arts Centre, Leicester. 21st September–16th December, 2018
Royal Albert Memorial Museum, Exeter. 19th January–16th March, 2019
Longside Gallery, Wakefield. 5th April–16th June, 2019
Southampton City Art Gallery, Southampton. 27th June–28th September, 2019

Journeys with 'The Waste Land'. 2018. Turner Contemporary. Curator Michael Tooby. (Group show).

Journeys with 'The Waste Land'. 2018. Coventry Cathedral and Herbert Gallery. (Group show).

Cultivation: Points of Vantage. 2017. Hestercombe Gallery. Curator, Tim Martin. (Group show).

Recent Acquisitions: New Art Gallery Walsall. Part 2: Landscape and the Urban Environment. 2017. (Group show).

Be Kind, Be Needed, Be Loved. 2017. The Barn, Banchory, Aberdeenshire. Final works of a three-year residency. Permanently sited in late 2017.

A Language from the Garden. 2017. Commissioned by the National Trust for Nymans gardens, West Sussex.

The Last Islands. 2016. Syson Gallery, Nottingham.

Soil Lab. 2016. 'The Dear Nature' letter installed as part of Newling's residency at the Barn, Banchory, Aberdeenshire.

Plant Culture. 2016. Arts Council Collection exhibition. Commissioned to make a new work, 'Birches and Berries'. Attenborough Gallery, Leicester.

Start Where You Are: Second Degree Potentias. 2016. Group exhibition for Bloc Projects, Sheffield.

The Last Islands and the Human Nature Tables. 2016. Inaugural exhibition for the new Syson Gallery, Nottingham.

The Lemon Tree and Me. 2016. live reading of the artist's book works. Transart, New York.

Periodic Tales, The Art of the Elements. 2015. Installation at Compton Verney, Warwickshire. Group exhibition.

Map of the Last Islands. 2015. Woodend Barn' Banchory, Aberdeenshire.

Magic Mountain. 2015. Museum of Contemporary Art, Santa Barbara, USA. (Group exhibition.

Generosity Plates. 2014. Wellcome Collection, London. Curated by James Peto.

One Foot in the Real World. 2013–2014. Irish Museum of Modern Art (IMMA), Dublin. Curated by Marguerite O'Molloy.

Market of Hidden Labours. 2013. Commissioned by York City Council. Managed by BEAM. Curated by Judit Bodor, A legacy commission for York.

Stall of Gathered Economies. 2013. New School House Gallery, York. Curated by Robert Teed and Paula Jackson.

Hedley Street: Notions of Value 'know you are loved'. 2013. 'Notions of Value' is a project by curator Kate Stoddart & John Newling together with Sherwood Rise Residents Association and funded by Arts Council England.

Ecologies of Value. 2013. Nottingham Contemporary. A retrospective of works from 1979–2012. Curated by Abi Spinks.

Lungs. 2012. Installation/performance. Invited artist for Contexts 2012 – The 2nd Sokolowsko Festival of Ephemeral Art. Curated by Malgosia Sady. Supported by the IN-SITU Foundation.

Moringa Oleifera with Seed. 2012. Invited artist, Royal Academy Summer Exhibition, London

Thomas Paine's Soil 'The Rights of Man'. 2011. Temple Gallery, Tyler School of Art, Philadelphia, USA. Curated by Robert Blackson. Soil constructed from 9 complete copies of 'The Rights of Man' text

The Miracle Trees. 2011. Commissioned installation. Nottingham Contemporary.

Singing Uncertainty. 2010. A live choral work involving the singing of questions written in hymns. The first live event took place at St Mary's Church, Nottingham.

Peterborough Soil (redistribution). 2010. This final aspect of the Peterborough Soil project saw the constructed soil added to the soil of the new Square.

The Clearing. 2010. (final phase). The nine beech trees from the laboratory phase are planted out as a permanent work. The Clearing will become a curated space for readings and research.

Synthia II (code/soil/life). 2010. This work was developed in collaboration with Bronislaw Szerszynski of the Department of Sociology and the ESRC Centre for the Economic and Social Aspects of Genomics (CESAGen), Lancaster University. Synthia II is a permanent work for Lancaster University.

Local History. 2010. Local history is a work developed from the Root Zone project in collaboration with Sherwood Library.

Root Zone. 2010. The Root Zone project is the cultivation of the area of land under the tree that is outside of our house. Root Zone is a work that gives me the opportunity to grow and harvest plants in a public space. These plants are used in particular projects. Currently the harvesting of these plants is involved in both Local History and Synthia II projects.

The Lemon Tree. 2010. The Lemon Tree is a living composition and has been acquired by the Collection for their contemporary art collection. The Collection is a Museum of contemporary art based in Lincoln.

What happens next is a secret. 2010. Installation at the Irish Museum of Modern Art. Group show.

Peterborough Soil. 2010. Installation at Peterborough Museum and Art Gallery.

Stamping Uncertainty. 2009. Installation in the Chapter House, Worcester Cathedral. Previously shown at Canterbury and Carlisle Cathedrals, amongst other places, this version of Stamping Uncertainty has been specially adapted for the Chapter House at Worcester Cathedral.

What We Do To Make Ourselves Feel Better. 2009. The penultimate public phase of Wellcome collection's commissioned project 'Make a Piano in Spain.' Funded by Wellcome Collection, the public space of Wellcome Trust.

Peterborough Soil. 2009-2010. Project commissioned by Opportunity Peterborough and Peterborough City Council and funded by EEDA with contributions from Peterborough Museum's arts development budget.

The Clearing (part 1). 2009. 'The Clearing', takes place in a laboratory housed in Bio-City Nottingham. A commissioned project by 'Hinterland'

Art in Public Places: The Archive Of The PADT (public art development trust). 2009. Proposals and drawings of Newling's projects from the archive exhibited. Leeds Art Gallery / Henry Moore Institute.

The Noah Laboratory (final phase) constructing soil. 2009. Installation and public transaction. Commissioned by 'The Collection' Lincoln.

A Whisper on a Public Square. 2009. Commissioned by and published on www.nottinghamvisualarts.net

Documents from the *Filter* series (1990). 2008. Exhibited as part of the permanent collection display at Leeds City Art Gallery.

Make a Piano in Spain. 2008. Installation. Part of the exhibition 'London Medicine', Wellcome Collection, London. Commissioned by Wellcome Collection, part of Wellcome Trust.

Out Loud. 2008. A live reading of the 500 citations collected from members of the public from 6th to the 10th of May, in answer to the question, 'What do you do to make yourself feel better?' Commissioned by the Wellcome Collection on behalf of the Wellcome Trust.

Ark: Road Vines. 2008–2012. 1st phase, Feasibility study and the setting up of the project's company Ark Lab. Ark lab will disseminate information on this major project whilst also acting as a trust fund organization.

'Prospects and Interiors: Sculptors drawings of inner space'. 2008. Three commissioned works for the Henry Moore Institute. Commissioned by the Henry Moore Institute. Group exhibition.

The Noah Laboratory. 2008–2009. Commissioned through the international residency program at the Lincoln collection.

Sacred in the Secular. 2007. Polish Museum of Contemporary Sacred Art, Kielce, Poland.

T*he Preston Market Mystery Project*. 2006–2007. Commission for Preston Market Place that involved the staging of three live events within the market site. Here + Now' commissioning group in conjunction with the Harris Museum, Preston.

The Westonbirt Wishes, Bronze Bolus (Final phase). 2006. Permanent installation of the wishes bolus on the site of the original collection of the wishes.

Chatham Vines. 2006. Final phase of the project. The Chatham Vines Wine was used in the Easter Liturgy at Rochester Cathedral as part of the dedicated Easter Eucharist.

Itch. 2006. a work exhibited from the 'Currency and belief' series, Chung-Ang University, South Korea. Group exhibition.

Stamping Uncertainty. 2006–2007. Grace Cathedral, San Francisco, USA

Stamping Uncertainty. 2006. The Fratry Building, Carlisle Cathedral.

Happiness is not a right. 2005. Performance / Installation, Cork, Ireland. European City of Culture

The Westonbirt Wishes (Phase 2). 2005. Book Launch and presentation, the National Arboretum, Westonbirt.

mine. 2005. Single version. St Mary Woolnoth, London

mine. 2005. Installation. Commissioned project, the Haggerty Museum of Art, Milwaukee, USA.

Transactions and Agreements. 2005. Multiple locations, Nottingham.

Stamping Uncertainty. 2004. Chapter House, Canterbury Cathedral.

International Environmental Artists Exhibition. 2004. Group Exhibition, Bradley University, Ohio, USA.

Chatham Vines. 2004–2006. Commissioned project, St Johns, Chatham.

... if spring didn't come. 2003. Phase 1, The Westonbirt Wishes. Commissioned project for Westonbirt National Arboretum.

Charity. 2003. Oxfam, Bold Street, Liverpool, Installation as part of Artranspennine03

A Cartography of Desire. 2003. Commission digital algorithm translating and converting

Currency and Belief. 2002–2003. 'A retrospective' Yorkshire Sculpture Park, England. Data from the lottery installation from the 'Currency and Belief' exhibition. Installed at www.john-newling.com

Shelter (1983). 'Sculpture in 20th Century Britain'. (Bought for the collection of Leeds City Art gallery and Museum 1984) Leeds City Art gallery and Museum. Group show.

Ices. 2003. Broadway Film Theatre, Nottingham, MADE, Birmingham.

Converge and Convey. 2003. Installation, 'Place' Castle Museum of Art, Nottingham. Group show.

Sleight of Hand. 2002. Performance, Yorkshire Sculpture Park.

Through the Surface. 2002. BWI galleries, Lublin, Poland. Group show.

Prayer Room. 2002. Commissioned permanent work for Worcester Acute Hospital.

1.0. 2001. Stafford Prison, Staffordshire.

Exhibitions (1991–2000)

Ices, Box. 2000. Angel Row Gallery, Commissioned by: The 'Box Project' Museum of Installation (London)
Aperture. 2000. Installation. Royal Mail Sorting Centre, Birmingham. Commissioned by: The Royal Mail
Immanent. 2000. Temporary installation. Victoria and Albert Museum, London. Commissioned by: Andrew Spira
Crossing. 1998. Temporary Installation. The Economist Plaza, London. Commissioned by: The Contemporary Arts Society.
Questions, Ices and Places. 1999. Commissioned by: The Laboratory at the Ruskin School of Drawing and Fine Art, University of Oxford
Outputs: Publication – Inserts project for 'City' journal, Volume 4, 2000
Physics of Place. 1998. Nottingham Castle. Commissioned by: Arts Council UK/ Lottery Funding
Ices. 1998. Temple Bar Gallery, Dublin. Commissioned by: Paul O'Neil and Ronan McCrea
Weight. 1998. Temporary Installation. Turnpike Gallery, Leigh. Commissioned by: Robert Hopper on behalf of Art Transpennine Company
Duration of A Wish. 1998. Permanent Installation. World Business Centre, Heathrow Airport. Commissioned by: B.A.A. Art Programme
Grail Text. 1997. Temporary Installation. A.J.Black, Nottingham Commissioned by: A.J. Black
Mad Hatter. 1996. Temporary Installation. Temple Bar, Dublin. Commissioned by: Temple Bar Developments of The Office of Public Works, Ireland.
Between (Even the Riddler Makes Wishes). 1996. Temporary Installation. Broadway Cinema, Nottingham. Commissioned by: Broadway Media Centre, Funded by the Foundation for Sport and the Arts.
The Place that is the site of the Inside of the Riddler's Jacket. 1995. Temporary Installation. Cornerhouse Manchester. Commissioned by: Stephen Snoddy
Where a Place becomes a site. 1995. Temporary Installation. Cornerhouse Manchester. Commissioned by Stephen Snoddy
Monotheism and Site. 1995. Temporary Installation. Cornerhouse Manchester. Commissioned by: Stephen Snoddy
Wishes. 1995. Temporary Installation. Cornerhouse Manchester. Commissioned by: Stephen Snoddy
100 places where Wishes and Petitions Meet. 1995. Temporary Installation. Cornerhouse Manchester. Commissioned by: Stephen Snoddy
Coated Wishes. 1995. Temporary Installation. Kilmainham, Dublin. Commissioned by: Artist Work Programme (AWP) at the Irish Museum of Modern Art
Exchange. 1995. Performance. Chapel at the Irish Museum of modern Art. Commissioned by: Artist Work Programme (AWP) at the Irish Museum of Modern Art
Growing Support: Walking Stick Cabbages. 1994. Commissioned by: Irish Museum of Modern Art
Mould. 1994. Permanent work, Sculpture. Inland Revenue, Nottingham. Commissioned by: Inland Revenue, Public Art Development Trust
Skeleton. 1994. Temporary Installation. All Saints Church, Newcastle Upon Tyne. Commissioned by: Locus +
Saturday Night, Sunday Morning. 1991. Market Square Nottingham Public Space. Commissioned by: Nottingham Arts Festival.
New Architecture. 1991. Sculpture, Long Row Gallery. Commissioned by: Angel Row Gallery
Lost. 1991. The Bonington Gallery, NTU. Installation. Commissioned by: The Bonington Gallery with support from the Henry Moore Trust

Exhibitions (1990–1979)

New Sculpture. 1990. Artsite Gallery, Bath (solo exhibition).
Filters. 1990. The Henry Moore Studio, Dean Clough, Halifax. Director: Robert Hopper. Commissioner: The Henry Moore Trust. Project Manager: Paul Bradley. Temporary installation. (solo exhibition).
John Newling. 1990. Edward Totah Gallery, London. (solo exhibition).
Invisible Cities. 1989. Leeds City Art Gallery. Group show.
1989. Feeringbury Manor, Colchester (group exhibition).
1989. Crammond Sculpture Park, Edinburgh. (group exhibition).
John Newling. 1989. Orchard Gallery, Derry, Northern Ireland. (solo exhibition).
John Newling. 1989. The Minories, Colchester (solo exhibition.
Recent Sculpture. 1988. Dean Clough Industrial Estate, Halifax. (solo exhibition).
John Newling. 1988. Edward Totah Gallery, London. (solo exhibition).
Sculpture. 1998. Krefelder Kunstverein, Krefeld, Germany. Invited as British representative in the Krefeld International Arts Festival. (solo exhibition).
Grey Matter New Sculpture '88. 1988. Ikon Gallery, Birmingham. (group exhibition).
Splitting the Nature and other works. 1987. Jollenbeck Gallery, Cologne.
New Sculpture. 1987. Edward Totah Gallery, London. (solo exhibition).
First Fulbright fellow in visual arts. 1985. Temporary works and public transactions made in numerous sites; including hotel rooms, a tourist so veneer stall, various streets, burnt-out cars, car parks and other spaces as places. Installations made in Los Angeles, San Francisco, Washington DC and New York
Ground Work Exhibition. 1984. Installation of 'Castle on skates with tent' on the archaeological site of the original Nottingham Castle. Outdoor sculpture, Nottingham Castle. (group exhibition)
New Sculpture; State, Church and other works. 1984. Midland Group, Nottingham. (solo exhibition)
New Works. 1983. Ian Birksted Gallery, London.
Working Drawings. 1983. Mineta Move Gallery, Brussels.
Art into the 80s. 1982. A group exhibition organized with Peter and Nigel Moores Foundation. Walker Art Gallery. Touring to the Fruitmarket Gallery, Edinburgh. And the Ikon Gallery, Birmingham. Participating artists; Glenys Barton, John Bellany, Gilbert & George, Harry Holland, John Hoyland, Ben Johnson, Kenneth Martin, Matta, Eammon McCabe, Henry Moore, John Newling, Victor Newsome, Ken Oliver and Michael Sandle. (group exhibition)
Sculptors' Drawings. 1982. Midland Group, Nottingham. (group exhibition)
'82 Group Show. 1982. Stoke on Trent Museum and Art Gallery. (group exhibition)
John Newling. 1982. Ian Birksted Galley, London. (group exhibition)
Sculptors. 1981. Chelsea School of Art. (group exhibition)
Seven Trestles. 1980. Midland Group, Nottingham. (solo exhibition)
The Nine Twists of Nature. 1979. Holt Street Gallery, Birmingham Arts Lab, Birmingham (solo exhibition)

Acknowledgments

We'd like to acknowledge the generosity and support of Fiona and Mark Hope, Jonathan Watkins and the team at Ikon Gallery for helping to make this project possible. Without their support, we would have not been able to realise the project at the timely juncture of the *Dear Nature* exhibition at Ikon. We would also like to thank Jeremy Webster, previously Director of The Collection, Lincoln, for commissioning the pivotal exhibition *The Noah Laboratory* and the essay *From My Garden* contained within the book, published to accompany the exhibition in 2009. Curators Alex Farquarson and Abi Spinks for the commissioning and careful curation of the *Ecologies of Value* exhibition at Nottingham Contemporary, which is a pivotal project within the book.

A further thank you is extended to Lorraine Grant and Simone Stewart at The Barn, Banchory, Sarah Shalgosky, Curator at the Mead Gallery, Mike Tooby, Trish Scott and the local audience that participated in the project at Turner Contemporary, Lisa Davies at the National Trust, Timandra Nichols and Judith King of Arts & Heritage and all the producers, installers and gallery teams that helped realise the projects in this book.

A special thank you to those that have contributed essays and writings – Anne Douglas and Mark Hope, Richard Davey and Jonathan Casciani. We would also like to thank Ann Newling for her continued support on every project and Reece Straw for his support on recent projects.

Beam Editions' Team
Project Editor and Creative Director – Jonathan Casciani
Design and Art Direction – Oliver Wood
Copy Editor – Aaron Juneau
Photography – James Gardiner, Stuart Whipps, Reece Straw

Ecology Works – John Newling

First edition published by Beam Editions, UK, 2020

ISBN 978-1-9162759-4-2

Contributing authors
Richard Davey
Ann Douglas
Mark Hope
Jonathan Casciani
John Newling
Jonathan Watkins

Editing
Aaron Juneau
Jonathan Casciani
Oliver Wood

Design and art direction
Oliver Wood
Jonathan Casciani

Photography
John Newling
James Gardiner
Reece Straw
Stuart Whipps
Graeme MacDonald (pp 104, 105)

Printed in the UK

www.beameditions.uk